Autonomous, Connected, Electric and Shared Vehicles

Disrupting the Automotive
and Mobility Sectors

Umar Zakir Abdul Hamid

Autonomous, Connected, Electric and Shared Vehicles

Disrupting the Automotive and Mobility Sectors

Autonomous, Connected, Electric and Shared Vehicles

Disrupting the Automotive and Mobility Sectors

BY

UMAR ZAKIR ABDUL HAMID, PHD

Warrendale, Pennsylvania, USA

400 Commonwealth Drive
Warrendale, PA 15096-0001 USA
E-mail: CustomerService@sae.org
Phone: 877-606-7323 (inside USA and Canada)
724-776-4970 (outside USA)
FAX: 724-776-0790

Library of Congress Catalog Number 2022947416
http://dx.doi.org/10.4271/9781468603484

ISBN-Print 978-1-4686-0347-7
ISBN-PDF 978-1-4686-0348-4
ISBN-epub 978-1-4686-0349-1

To purchase bulk quantities, please contact: SAE Customer Service

E-mail: CustomerService@sae.org
Phone: 877-606-7323 (inside USA and Canada)
724-776-4970 (outside USA)
Fax: 724-776-0790

Visit the SAE International Bookstore at books.sae.org

Publisher
Sherry Dickinson Nigam

Development Editor
Publishers Solutions, LLC
Albany, NY

Director of Content Management
Kelli Zilko

Production and Manufacturing Associate
Brandon Joy

Table of Contents

Foreword by Dr. Rahul Razdan   xiv

Foreword by Daniel Watzenig   xvi

Preface and Acknowledgments   xviii

SECTION 1 — Introduction

CHAPTER 1

Introduction   3

1.1.   FIR and the Automotive Industry   3

1.2.   ACES as the Future Mobility   4

1.3.   Objectives of the Book   6

1.4.   Expectation and Contribution of the Book   7

1.5.   Overview and Outline   8

1.6.   Summary   9

References   9

SECTION 2 — ACES as the Future Mobility: Background

CHAPTER 2

Recent Events and Progress
Propelling ACES Growth   15

2.1.   Advanced Driver Assistance Systems and
       Vehicle Automation   16

2.2.   Active Safety and the Safety Benefits   17

2.3.   Software-Defined Vehicles   18

2.4.   Fourth Industrial Revolution   19

2.5.   United Nations Sustainable Development Goals 19

2.6.   Society 5.0   20

2.7.   Climate Change and Intergovernmental
       Panel on Climate Change 2021 Report   20

2.8.   Carbon-Neutral and Finite Petroleum
       Resources   22

2.9.   Green, Circular, and Sustainable Economy   23

2.10. Sharing Economy   23

2.11. Regulations Facilitating ACES   24

2.12. IoT and 5G   24

2.13. Advances in Computing Power and Platforms   25

2.14. Sensor Advancements (LiDAR, RADAR,
       Camera)   26

2.15. Cross-Functional Software Product
       Development and Change Management   28

2.16. Silicon Valley and Rise of Start-Ups   28

2.17. X-as-a-Service   29

2.18. Acqui-Hiring and M&A   30

2.19. Supply Chain Evolutions   31

2.20. Digital Natives   31

2.21. Summary   32

References   32

SECTION 3 — Concise Overview of ACES

CHAPTER 3

Autonomous Vehicles: Concise
Overview   41

3.1. Background and What   41

3.2. Technical Overview   43

    3.2.1. How Humans Operate: An Anecdote to
         Simplify the AV Back-End Algorithms   43

    3.2.2. Human-Operated Vehicles Operation   44

    3.2.3. AV Stack in General   46

    3.2.4. Mapping and Localization   47

    3.2.5. Environmental Awareness   48

    3.2.6. Risk Assessment   50

    3.2.7. Motion Planning   52

    3.2.8. Motion Control   54

    3.2.9. Interface, Calibration, and Monitoring   55

3.3. Vision and Current State of the AV Industry   56

3.4. Summary   58

References   58

CHAPTER 4

Connected Vehicles: Concise Overview   65

4.1. Background   65

4.2. Connecting and Connected   67

4.3. Back-End Overview   68

    4.3.1. Vehicular Ad Hoc Network   68

    4.3.2. Cloud Technology   69

4.3.3. Dedicated Short-Range Communications	69
4.3.4. Cellular Vehicle-to-Everything (C-V2X)	70
4.3.5. Low-Power Wide-Area Network	71
4.3.6. 5G and Its Importance for Vehicle Connectivity	71
4.3.7. A Glimpse of 6G, the Probable Next Step in the Vehicle Connectivity Field	72
4.4. Applications	**73**
4.4.1. Vehicle-to-Everything	73
4.4.2. Vehicle Connectivity for Improved Active Safety	74
4.4.3. Vehicle Platooning	75
4.4.4. Improved Infotainment	76
4.4.5. Improved GPS and Traffic Jam Reduction	76
4.4.6. Vehicle Connectivity Enabling Automated Delivery	77
4.4.7. Vehicle Connectivity Improving Shared Mobility	77
4.5. Vehicle Connectivity Roles in Enabling True ACES Mobility	**78**
4.6. Vision and Current State of the CV Industry	**79**
4.7. Summary	**80**
References	**80**

CHAPTER 5

Electric Vehicles: Concise Overview   87

5.1. Background	**87**
5.2. Electrifying the World—The Motivation	**90**
5.3. Back-End Overview	**92**
5.3.1. Hybrid Electric Vehicles	92
5.3.2. Plug-In Hybrid Electric Vehicles	95
5.3.3. Battery Electric Vehicles	96
5.3.4. Fuel Cell Electric Vehicle	98
5.3.5. Solar EVs	100

5.4.   Vehicle Electrification Roles in Enabling
        True ACES Mobility   101

5.5.   Vision and Current State of the EV Industry   101

5.6.   Summary   102

References   102

CHAPTER 6

Shared Mobility: Concise Overview   107

6.1.   Background   107

6.2.   Motivations behind Shared Mobility   108
        6.2.1. Sharing Economy   108
        6.2.2. Uberization   110
        6.2.3. X-as-a-Service   110

6.3.   Terminology and Definition: Clarification
        and Difference   111
        6.3.1. Shared Mobility   112
        6.3.2. Carsharing   112
        6.3.3. Carpooling   112
        6.3.4. Ridesharing   113
        6.3.5. Ridehailing   114
        6.3.6. Ridesourcing   114
        6.3.7. Micromobility   114
        6.3.8. Paratransit   115
        6.3.9. Microtransit   115
        6.3.10. Other Types of Shared Mobility   116

6.4.   Differences between "Traditional Taxi" and
        "Shared Mobility"   116

6.5.   Customer-and-User-Facing Technology
        Development for Shared Mobility   116

6.6.   Roles of Shared Mobility in the ACES
        Ecosystem   117

6.7.   Summary   119

References   120

SECTION 4 — Disruptions, Challenges, and Benefits of ACES

CHAPTER 7

Disruptions Caused by ACES
Mobility   125

7.1.	Background	125
7.2.	Social Disruptions	125
	7.2.1. Work-Life Balance Improvements	126
	7.2.2. Disruptions for Media and Infotainment	126
	7.2.3. Merging with the Other Industrial Revolution 4.0 Ongoing Developments	128
	7.2.4. Redefining Mobility	128
	7.2.5. Disrupting Future Urban Planning	128
7.3.	Legal, Economical, and Workforce Disruptions	130
	7.3.1. Insurance Industry	130
	7.3.2. Taxations	130
	7.3.3. New Required Skills for Workforce	131
	7.3.4. New Job Ecosystems Opportunities	132
	7.3.5. Who Will Have the Ownership of ACES Vehicles?	132
	7.3.6. Indirect Influence on Popular Culture	132
	7.3.7. Passenger Behavior during the ACES Journey	133
	7.3.8. Regulations... and More Regulations	133
	7.3.9. Private-Public-People Partnerships	134
	7.3.10. Increased Requirements of Empathy from the Leaders to Employees	135
7.4.	Technical, Technological, and Industrial Disruptions	135
	7.4.1. ACES Increases the Software Importance in the Automotive and Mobility Industry	136
	7.4.2. Changes in the Business Models	136
	7.4.3. Process and Regulations Changes	136
	7.4.4. Infrastructural Disruptions	137
	7.4.5. New Automotive and Mobility Industrial Stakeholders	138

7.4.6. Democratizing Mobility   138
7.4.7. New Incomes for Countries   138
7.4.8. Skunkworks Project-Organizations   138
7.4.9. More Concentrated Efforts to Tackle
       Climate Change Issues   139

7.5. Summary   **139**

References   **140**

CHAPTER 8

Potential Challenges of ACES   145

8.1.   **Background**   **145**
8.2.   **Technical Challenges of ACES**   **146**
       8.2.1. Dealing with Uncertainties   146
       8.2.2. Network Latency   147
       8.2.3. Cybersecurity   147
       8.2.4. Range Anxiety Is Still an Issue   147
       8.2.5. Infrastructural Improvements to Support
              Vehicle Electrification   148
       8.2.6. Battery Recycling and Waste Management   148
       8.2.7. Hygiene Topics   148
       8.2.8. Ridesharing and Traffic Jams   149
       8.2.9. Traveling Salesman Problem   150

8.3.   **Legal, Industrial, and Workforce Challenges
       of ACES**   **150**
       8.3.1. Vandalism and Petty Crimes   150
       8.3.2. Whose Fault Is It? Who Is to be Blamed?   150
       8.3.3. Scope Creep and Technical Debt in the
              Software Development   151
       8.3.4. Pricing Is Still Expensive   151
       8.3.5. Unclear Requirements because of the
              Knowledge Gap in the Business-Facing
              Organizations   152

8.4.   **Social and Ethical Challenges of ACES**   **153**
       8.4.1. Changing Job Landscapes   153
       8.4.2. Transparency Is Needed   153

8.4.3. Requirement for a Visionary and Skillful
Public Leadership   154

8.4.4. Importance of Education and Propagations   154

8.4.5. False and Misleading Marketing   154

8.5.   **Summary**   **155**

References   **156**

CHAPTER 9

Potential Benefits of ACES   159

9.1.   **Technological, Safety, and Security Benefits**   **159**

9.1.1. Prompting Safer Automation in Other
Industries   160

9.1.2. Encouraging ACES Mobility across
Different Transportation Domains   160

9.1.3. Increased Safety   161

9.1.4. More Transparency and Security for Mobility   163

9.2.   **Societal and Sustainability Benefits**   **164**

9.2.1. Lessened Car Ownership: A Chance to
Reimagine Mobility   164

9.2.2. Cleaner Mobility Ecosystem and Energy   165

9.2.3. Improved Traffic in Cities   165

9.2.4. More Spaces for Urban Recreational
Zones and Activities: Improving the City
Attractiveness   166

9.2.5. Less Stress for City Dwellers   167

9.2.6. Potential to Reduce Crimes   167

9.2.7. Better Social Security Benefits   168

9.2.8. Facilitating Some Objectives of the UN SDGs 169

9.2.9. Improving Air Quality, Improving Health   170

9.3.   **Mobility User Benefits**   **171**

9.3.1. Better Transportations for the Society   171

9.3.2. Reduced Costs of Transportation   172

9.4.   **Economic Benefits**   **172**

9.4.1. New Investments Potential   172

9.4.2. Opportunities for Developing Countries   173

9.4.3. New Business Models   173

9.4.4. Cross-Collaborations between Different
Industries   174

9.5. Summary   174

References   174

SECTION 5 — Summary and Conclusions

CHAPTER 10

Summary: "ACES Is Imminent. It Is a Bumpy Road. Cross-Organizational Collaborations Are a Necessity."   179

10.1. Recapitulating the Book   180

10.2. Summary for Section 1—Introduction   181

10.3. Summary for Section 2—ACES as the
Future Mobility: Background   181

10.4. Summary for Section 3—Concise Overview
of ACES   182

10.5. Summary for Section 4—Disruptions,
Challenges, and Benefits of ACES   182

10.6. Required Future Efforts for ACES Mass
Deployment   183
10.6.1. Technical   183
10.6.2. Legal   183
10.6.3. Social   184
10.6.4. Process and Procedures   185

10.7. It Is a Bumpy Road, But It Is Not Impossible   185

10.8. The Dream and Hope for This Book   186

10.9. Summary and Conclusion   187

References   187

Index   189
About the Author   193

Foreword

By Dr. Rahul Razdan
CEO Razdan Research Institute

Historically, electronic solutions driven by semiconductors have fundamentally shifted major parts of the world economy through major technology waves (Figure 1). The first wave consisted of the age of centralized computing and the leaders in the field which included companies such as IBM, Digital Equipment Corporation, Wang, and others. These technologies provided productivity solutions for the administrative (G&A) functions of the global business

FIGURE 1 Megatrends driven by electronics design.

Economy Shaping Technologies...

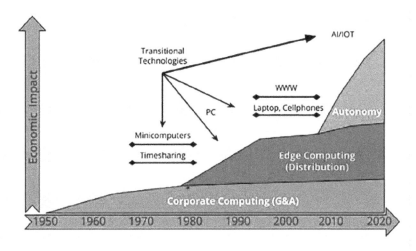

enterprise. With this shift, the finance, human resources, and administrative functions of global business were disruptively impacted. Gone were the days of a sea of administrators doing paperwork.

The next wave consisted of edge computing devices (color red in Figure 1) such as personal computers, cell phones, and tablets. With this capability, companies such as Apple, Amazon, Facebook, Google, and others added enormous productivity to the advertising and distribution functions for global businesses. Suddenly, one could directly reach any customer anywhere in the world. This megatrend has fundamentally disrupted markets such as education (online), retail (e-commerce), entertainment (streaming), commercial real estate (virtualization), health (telemedicine), and more.

Today, we are at the beginning of the next major disruptive cycle caused by computing. This cycle consists of embedded sensory devices (sometimes known as the Internet of Things), local intelligence systems (sometimes known as machine learning), and global intelligence (sometimes known as cloud resources). Broadly called AI/IoT, these three technologies will disruptively impact nearly every market segment where in-field sensing with computing can solve interesting problems. Medical, mining, agriculture (land or ocean), space operations, and, of course, transportation are examples.

In transportation, the term Automated, Connected, Electric, and Shared (ACES) has been coined to represent the enormous innovations enabled by the underlying electronics technology. The previous megatrends have fundamentally reshaped society, and this next wave promises to do the same. This book outlines the ecosystem around ACES, the underlying technology drivers, and the implications of this paradigm shift. Given the potential for transformational change, this content is important not only for transportation specialists but also for the broader society.

Dr. Rahul Razdan

CEO Razdan Research Institute (www.razinstitute.com)
Senior Director, Special Projects, Florida
Polytechnic University

Foreword

By Daniel Watzenig
Professor at Graz University of Technology, Austria

Electrification, automatization, digitalization, and standardization are four major trends dominating our society, especially toward the goal of green and sustainable mobility. Therefore, autonomous, connected, electric and shared vehicles (ACES) are destined to be game changers on the roads, altering the face of mobility as we experience it by today. The expected benefits coming from ACES vehicles range from improved safety, reduced congestion and lower stress for car occupants, social inclusion, lower emissions, to better road utilization because of optimal integration of private and public transport.

These vehicles will be able to exchange data and communicate with other vehicles and objects in the immediate surroundings. As the mobility industry is rapidly deploying new forms of carsharing, ridesharing, and ridehailing services, the importance of private ownership of vehicles will be reduced. Shared mobility services are also taking steps to fully integrate into public transportation systems to offer seamless multimodal experiences. These are all intended to happen with fully electric vehicles to ensure the reduction of emissions in mobility.

However, despite tremendous improvements in sensor technology, high-performance computing, machine learning, computer vision, data fusion techniques, control system design, communication bandwidth (in-vehicle Ethernet, V2X…), and other system technology areas, the market introduction of electric, connected, and shared vehicles that are fully automated and capable of unsupervised driving in an unstructured environment still remains a long-term goal.

Beyond the technological issues, several regulatory action items for faster introduction of automated vehicles still have to be resolved by the governments to ensure full compatibility with the public expectations regarding legal responsibility, safety, and privacy. Authorities need to create the legal framework to remove liability traps to encourage test regions, promote long-term infrastructure investments, provide open access, and set up legal frameworks for V2V communication.

Of course, the many challenges that still need to be solved to drive this exciting development forward have motivated the creation of this book. All chapters of the book have been carefully selected to cover certain aspects of the development and deployment of ACES vehicles.

We hope that the reader will be inspired by the different chapters and gain insight into ACES as the future of mobility.

Daniel Watzenig
Professor at Graz University of Technology, Austria
Head of Department at Virtual Vehicle Research
Editor-in-Chief of the *SAE International Journal of Connected and Automated Vehicles* (JCAV)
Lecturer at Stanford University

Preface and Acknowledgments

After more than a year of preparing the contents for the book, finally, we have reached this part, i.e., composing the preface and the acknowledgment.

This book is written because I believe, based on my eight-year journey in the Autonomous, Connected, Electric, and Shared (ACES) Mobility field, ACES will not only disrupt the automotive and transportation sectors, it will also transform the job market landscape. This will surely directly impact and affect the younger generations. However, with the media hype surrounding the technology, it is easy to misinterpret the technology and miss the big picture. Therefore, I am expecting this book to serve this purpose, i.e., to provide a general overview of ACES technology for cross-disciplinary audiences.

There are many people who I would like to thank for encouraging me to complete the book; however, I believe there is not enough space for everyone (and I have thanked them individually in person). Accordingly, I would like to give special shout-outs to the SAE publication team which includes Sherry Nigam and Linda DeMasi. Finally, a special appreciation to Nora for the support during the final part of the publication.

Umar Zakir Abdul Hamid, PhD
2021–2022

SECTION 1
Introduction

Introduction

The advancement of science and technology has always brought with it revolutionary elements. In the past few centuries, humankind has seen multiple industrial revolutions occur. In the present day, as the world is changing faster and becoming more borderless, the spread of knowledge and talents have become easier. Progress has facilitated the birth of innovations such as the Internet, which is extremely helpful in disseminating information that was previously only accessible in certain countries. This has sparked a new kind of industrial revolution, known as the Fourth Industrial Revolution (FIR), which encompasses different sectors, including automotive.

1.1. FIR and the Automotive Industry

The FIR has demonstrated a massive disruption potential in the automotive and mobility sectors [1, 2, 3]. With the advances in computational platforms and hardware devices, vehicle automation topics have been flourishing, especially in the past few years. According to reports, Autonomous Vehicles (AVs) are expected to be worth more than $7 trillion by 2050 [4]. However, the automotive and mobility disruption is not only caused by AVs. Instead, AV is in reality the last piece of the puzzle that will complete a larger concept of future mobility, which should be preceded by vehicle electrification and connectivity topics as well as shared mobility [5]. Together all elements are creating a new automotive and industrial ecosystem that will disrupt not only the technical development but also the management and supply chain of the industry, among many others.

The FIR is not only bringing new technology, but it is also revolutionizing and redefining the automotive industry. The previously mechanical-laden industry is now heavily scouting for software technical expertise [6]. The switches are also impacted by several external factors such as finite petroleum resources, which encourage electrification and automation across different transportation and automotive sectors [7]. All these trends are leading humankind toward future mobility which will be Autonomous, Connected, Electric, and Shared (ACES).

1.2. ACES as the Future Mobility

The rise of next-generation vehicle companies such as Tesla and NIO, among many others, did not happen overnight [8]. Behind these growths lie four different topics of ongoing research works on revolutionary technologies, i.e.:

- Vehicle Automation
- Vehicle Electrification
- Vehicle Connectivity
- Shared Mobility

If analyzed separately, all four elements as their standalone elements have a smaller and lesser impact. However, when combined and discussed as a whole, they form a disruptive ecosystem. Billions have been invested in the last decade alone in this matter, which highlights the importance of this topic [9] (Figure 1.1). In addition, with the advent of connectivity and fifth-generation (5G) technology, the mobility concept is also being redefined. The discussions eventually reached the reduced vehicle ownerships topic with the progress made by companies such as Uber in the shared mobility area [10].

Being wary of these developments, countries all around the world started to produce blueprints, discussions, and studies to comprehend the phenomenon. Pilot projects are being operated all around the world to give trial platforms for not only traditional automotive carmakers but also sandbox programs of deep technology companies to pilot their ACES-related technologies [11, 12, 13]. KPMG (Klynveld

FIGURE 1.1 The total investments in 2021 for ACES technology since 2010 [9].

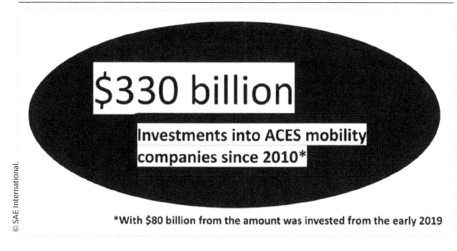

© SAE International.

*With $80 billion from the amount was invested from the early 2019

Peat Marwick Goerdeler) International Limited has reported a detailed yearly index of AV Readiness every year by different countries globally, which highlights the "race" among different nations in preparing their community for the disruptive technology [14, 15, 16]. Regulations have also been prepared for the arrival of ACES. For example, Germany has released a proposal for a new regulation with regard to Autonomous Driving (AD) in late 2020 [17]. Singapore is another country to do this, where in 2021, the nation created the National Technical Reference for Autonomous Vehicles [18].

Thus, it is not shocking when the World Economic Forum mentioned that, in conjunction with the FIR, 65% of the future jobs do not exist yet [19]. With the ACES, the effects can be already seen today as companies are starting to hire more software engineers in addition to the traditional mechanical engineering, which is also getting more influenced by the software paradigm.

However, despite the growth, among the non-technical and nonautomotive public audience the discussions are still limited, and people are still relying on the overhyped perspective on ACES from the mass media, which sometimes overshadows the real status and actual progress of the work. This is creating a conflict between technical and business perspectives—which if not handled properly can prompt the delivery of unreliable and unsafe automotive ACES software.

As ACES technical product development works are ongoing, to make sure that the user acceptance toward the technology is parallelly good and the ethical usage by the end users are assured, the non-technical and nonautomotive people should also grasp the ACES concept, at least to a minimum level. In addition, for automotive practitioners, as ACES growth is ongoing and at a remarkably fast rate, a lot of the discussions happen only in a limited spectrum and rarely on the big picture of the ACES concept. This has created a legacy silo organization that delays the progress of mass productization of ACES. Thus, centralized and focused discussions should be arranged along with constant education to ensure a centralized understanding of the big picture.

1.3. Objectives of the Book

The disruptions of the industry are prompted by not only the technology but also a shift between traditional and software-based mindsets, which also involve the arrival of new-generation workforces in the automotive industry. According to Kreuzer and Tomaschitz [20], one of the most important things that are often undervalued by industry when talking about the productization of new emerging technologies is the necessity of organizational transformation. The transformation demands continuous education to facilitate new types of management and operation methods. Failure to address this topic will likely cause the industry to provide faulty products and operations. In addition, when dealing with emerging technologies with disruptive impacts, organizations should learn from the history of the demise of corporations in the previous wave of digitalization in the past decade. The failure to learn, be nimble, and be agile will accelerate the collapse of even a giant business. However, as the developments are too rapid, the whole ecosystem is not clearly understood yet even by the practitioners in the automotive field as it is a work in progress.

The rate the industry is evolving is extremely staggering [21]. Recently, a lot of companies are consolidating and making "mergers and acquisitions" (M&A) on the topic of ACES, which proves that something big is brewing in the industry [22]. Furthermore, the

FIGURE 1.2 ACES has a high potential of disrupting how humans will live in the future.

Solveig Been/Shutterstock.com.

changes will also bring societal consequences (e.g., career landscape evolutions, potential new regulations, and how humans will live). For example, Society 5.0 concepts discussed the way human society will live when the FIR technology entered the market en masse [23] (Figure 1.2). Despite the many potential benefits to society, the extent of disruption that will be caused by ACES is still a myriad of uncertainties. Thus, efforts are needed to promote a better understanding of this topic by different parties.

Therefore, this book is written to educate not only the practitioners but also the general public, as well as experts from multidisciplinary fields related to the automotive and mobility ecosystems, as a one-stop center for knowledge dissemination on ACES. By reading this book, the author is hoping that the audience will get an overview of the future mobility concept through the descriptions of ACES Vehicles. This book will help bridge the gap among the different perspectives of the audience.

1.4. Expectation and Contribution of the Book

Based on the discussions in the previous section, the author writes this book with the intention of providing an overview on future mobility and ACES technology. As such, the book will cover these topics:

- Recent trends and progress stimulating ACES progress
- ACES overview
- Disruptions of ACES to the automotive and mobility industry
- Challenges of ACES implementation
- Potential benefit of ACES ecosystem

This book is a conscious personal effort by the author to discuss and provide coverage on the disruptive impacts brought soon by ACES vehicles on the automotive and mobility sectors. As such, the author evaluates the impact that will be brought by ACES technology individually and as an ecosystem. This in return will be beneficial to a vast number of readers who are interested in the topics of AVs, Connected Vehicles (CVs), Electric Vehicles (EVs), and Shared Mobility.

The author strongly believes that this book will attract the interest of researchers, academics, and professionals from different viewpoints (i.e., automotive, mobility, social sciences, law, and safety, among many others). It is worth noting that this book does not aim to provide extensive technical details on ACES but instead to enable people who are new to the topic to fully understand the system. In addition, by reading this book, readers will also be able to understand there will be an evolution within the automotive business and industrial landscape.

1.5. Overview and Outline

To ensure that the book can be read as standalone chapters and coherently from cover to cover, the book is divided into five sections. In the second section (Chapter 2), following this Introduction chapter, the author provides an overview of the future mobility background, which involves a survey on the recent events and progress which drive and stimulate the productization of ACES. In the third section (Chapters 3–6), the author provides a technical overview of each of the ACES elements in their respective chapters. By reading each chapter, the reader will be provided with a comprehensive indicative survey on each component of ACES. The disruptions, challenges, and benefits of ACES are then discussed separately in the fourth section (Chapters 7–9). Finally, in the final section (Chapter 10), the author deduced the conclusions of the book as well as some future work suggestions.

1.6. Summary

In this chapter, the author brings to the attention of the readers a brief background, the objectives, and key expectations of the book. To recap, this book aims to provide a comprehensive overview of ACES technology as well as the future of mobility. The author recommends that the book be read from cover to cover; however, to allow speed reading by some readers, each chapter is written in a way that can be read as a standalone chapter. The author wishes the readers well, and as this book is intended for casual introductory reading with very broad perspectives, the author strongly encourages readers to use the cited list of references for further reading. It is again worth noting that this book is not intended to be a technical textbook, but instead, the aim is to provide an indicative overview of the future mobility trends rather than being technically exhaustive.

References

1. Pardi, T., "Fourth Industrial Revolution Concepts in the Automotive Sector: Performativity, Work and Employment," *Journal of Industrial and Business Economics* 46, no. 3 (2019): 379-389.

2. Armengaud, E., Sams, C., Von Falck, G., List, G. et al., "Industry 4.0 as Digitalization over the Entire Product Lifecycle: Opportunities in the Automotive Domain," in Stolfa, J., Stolfa, S., O'Connor, R., and Messnarz, R. (eds.), *European Conference on Software Process Improvement* (Cham: Springer, 2017), 334-351.

3. Lin, D., Lee, C.K.M., Lau, H., and Yang, Y., "Strategic Response to Industry 4.0: An Empirical Investigation on the Chinese Automotive Industry," *Industrial Management & Data Systems* 118, no. 3 (2018): 589-605.

4. Morris, D.Z., "Driverless Cars Will Be Part of a $7 Trillion Market by 2050," *Fortune*, June 3, 2017.

5. Ehlers, C., "Mobility of the Future-Connected, Autonomous, Shared, Electric," in *30th International AVL Conference "Engine & Environment"*, Graz, Austria, 175-178, 2018.

6. Porzelt, J., "Strategic Impacts on Automotive Incumbents in the Market of Autonomous Driving," PhD dissertation, Universidade Católica Portuguesa, 2020.

7. Salah, W.A., Albreem, M.A.M., Alsayid, B., Zneid, B.A. et al., "Electric Vehicle Technology Impacts on Energy," *International Journal of Power Electronics and Drive Systems* 10, no. 1 (2019): 1.

8. Teece, D.J., "Tesla and the Reshaping of the Auto Industry," *Management and Organization Review* 14, no. 3 (2018): 501-512.

9. Holland-Letz, D., Kässer, M., Kloss, B., and Müller, T., "Mobility's Future: An Investment Reality Check," McKinsey & Company, May 27, 2021, accessed August 9, 2022, https://www.mckinsey.com/industries/automotive-and-assembly/our-insights/mobilitys-future-an-investment-reality-check

10. Ward, J.W., Michalek, J.J., Azevedo, I.L., Samaras, C. et al., "Effects of On-Demand Ridesourcing on Vehicle Ownership, Fuel Consumption, Vehicle Miles Traveled, and Emissions per Capita in US States," *Transportation Research Part C: Emerging Technologies* 108 (2019): 289-301.

11. Hamid, U.Z.A., Ishak, S.Z., and Imaduddin, F., "Current Landscape of the Automotive Field in the ASEAN Region: Case Study of Singapore, Malaysia and Indonesia—A Brief Overview," *Asean Journal of Automotive Technology* 1, no. 1 (2019): 21-28.

12. Hafiz, D. and Zohdy, I., "The City Adaptation to the Autonomous Vehicles Implementation: Reimagining the Dubai City of Tomorrow," in Hamid, U.Z.A. and Al-Turjman, F. (eds.), *Towards Connected and Autonomous Vehicle Highways* (Cham: Springer, 2021), 27-41.

13. Ainsalu, J., Arffman, V., Bellone, M., Ellner, M. et al., "State of the Art of Automated Buses," *Sustainability* 10, no. 9 (2018): 3118.

14. Threlfall, R., "Autonomous Vehicles Readiness Index," Klynveld Peat Marwick Goerdeler (KPMG) International, 2018.

15. KPMG, "Autonomous Vehicles Readiness Index—KPMG Global," 2019, accessed August 9, 2022, https://home.kpmg/xx/en/home/insights/2019/02/2019-autonomous-vehicles-readiness-index.html

16. Threlfall, R., "Autonomous Vehicles Readiness Index (AVRI)—KPMG Global," 2020, accessed August 9, 2022, https://home.kpmg/xx/en/home/insights/2020/06/autonomous-vehicles-readiness-index.html#download

17. Publisher, "Deutschland wird international die Nummer 1 beim autonomen Fahren," n.d., accessed August 9, 2022, https://www.bmvi.de/SharedDocs/DE/Artikel/DG/gesetz-zum-autonomen-fahren.html

18. Toppan Leefung Pte. Ltd., "Technical Reference for Autonomous Vehicles," Singapore, accessed July 2022, https://www.singaporestandardseshop.sg/Product/SSPdtPackage/063be487-9a8b-40af-995e-2aac06d5516a

19. McLeod, S. and Fisch, K., "Chapter 1: The Future of Jobs and Skills," World Economic Forum, 2016, https://reports.weforum.org/future-of-jobs-2016/chapter-1-the-future-of-jobs-and-skills/

20. Kreuzer, V.I. and Tomaschitz, M., "Organizational Challenges in Automotive Development," in Hick, H., Küpper, K., and Sorger, H. (eds.), *Systems Engineering for Automotive Powertrain Development* (Cham: Springer International Publishing, 2021), 123-146.

21. Ebert, C. and Favaro, J., "Automotive Software," *IEEE Software* 34, no. 03 (2017): 33-39.

22. Alkaraan, F., "Strategic Investment Decision-Making: Mergers and Acquisitions toward Industry 4.0," in Cooper, C.L. and Finkelstein, S. (eds.), *Advances in Mergers and Acquisitions* (Bingley, UK: Emerald Publishing Limited, 2021), doi:10.1108/S1479-361X20210000020004

23. Deguchi, A., Hirai, C., Matsuoka, H., Nakano, T. et al., "What Is Society 5.0," *Society 5* (2020): 1-23.

SECTION 2

ACES as the Future Mobility: Background

Recent Events and Progress Propelling ACES Growth

The discussions on ACES mobility do not occur overnight. It is an avalanche result of continuous developments by researchers and practitioners in the past decades. Despite some elements of ACES being heavily riddled with "hype" by the mass media, there are plenty of evidence, trends, and advances which support and justify the ACES claim as the future mobility. Unfortunately, most of these trends and advances are reported in various media outlets in different fields. The author strongly believes that understanding the recent events and progress which propelled ACES growth is important to allow better comprehension of the current direction of the automotive industry. In addition, this knowledge will also guide experts and practitioners in producing future mobility creations and offerings which are useful and desired by the end users and customers. Consequently, in this chapter, the author identifies 20 trends and advances which are deemed to be contributing to the progress of the growth of future mobility. The chapter is written with indicative and suggestive writing concepts, i.e., to provide an overview of the ACES growth factors, instead of aiming to be an exhaustive survey. The author believes this chapter will help the new audience become quickly familiar with ACES topics. It will also guide the readers of this book to comprehend better the ensuing discussions.

2.1. Advanced Driver Assistance Systems and Vehicle Automation

In 2015, William Fleming has outlined in an extensive article the progression of the automotive industry in the past forty years. Fleming divided the review based on five categories, which include Infotainment, Advanced Driver Assistance Systems (ADAS), Automotive Body Electronics, Automotive Chassis Electronics, and Automotive Powertrain Electronics. It is clearly stated from the review how far the concept of mobility has changed from just being a transportation platform to now also increasingly addressing lifestyle and leisure demands [1]. ADAS have been one of the stepwise enablers toward more vehicle automation productization. With the milestones achieved by ADAS, vehicle automation has been getting more attention from researchers and engineers. Since 2010, more personal vehicles are manufactured equipped with the technology, for example, the introduction of City Safety by Volvo as well as VW Front Assist from Volkswagen [2] (Figure 2.1). The growth of the said innovations was facilitated by the continuously evolving safety regulation requirements. For example, in the

FIGURE 2.1 Automation helped to improve the vehicle safety.

Pepermpron/Shutterstock.com.

European Union, active safety systems such as Autonomous Emergency Braking (AEB) and intelligent speed assistance are required by law to be established in new vehicles [3]. The growth of vehicle automation also indirectly encouraged the foundation of the New Car Assessment Program for Southeast Asian Countries (ASEAN NCAP) in the Southeast Asian region [4]. As one of the regions with the highest road fatalities globally, offering ADAS in the ASEAN market has helped to increase the public and user attention to vehicle automation technology [5]. As a result, following the ADAS progress in the market, car manufacturing companies are now attempting to provide more vehicle automation technologies in their products, which has served as one of the catalysts toward the development of ACES technology, particularly in connected and autonomous mobility topics.

2.2. Active Safety and the Safety Benefits

One of the strongest justifications to bring ACES into the market is because of the vast advantages and future potential benefits that the technology will bring to vehicle safety [6]. Increased elements of vehicle automation technology have reached the market via vehicle active safety features. For example, technologies such as AEB, Adaptive Cruise Control, and Emergency Lane Keeping have been in the market for a while. An extensive study highlighted that the adoption of vehicle automation technology such as reversing cameras has significantly reduced the number of backover injury rates [7]. Furthermore, each of the technologies in ACES has the vast possibility of safety on its own. For example, in Hamid, Abdul, and Limbu [8], it is revealed that by incorporating vehicle connectivity technology, the vehicles can transmit information to each other if anomalies or roadwork are happening in the traffic in front of them. This has the potential to reduce accidents as well as traffic jam occurrences, indirectly motivating the growth of ACES productization.

2.3. Software-Defined Vehicles

This might be a repetition of the previous two factors mentioned in Subsections 2.1 and 2.2. However, the term *Software-Defined Vehicle* as one of the factors propelling ACES growth is worth mentioning here (Figure 2.2). It was reported that the automotive software industry was worth 13.1 billion US dollars (USD) in 2019 alone [9]. According to a report by Deloitte, a professional services network, "software-defined vehicle" refers to the increasing value of the software in a vehicle compared to its hardware [10]. This eventually foresees that in the future, a vehicle worth will more probably be determined by the software that it will offer compared to the hardware offering. And it is not surprising that a lot of automotive software start-ups were founded in the last decade. With major automotive carmakers like Volkswagen creating a subsidiary organization totally focusing on software (CARIAD), it signals that a software-defined vehicle is not just mere hype. This paradigm eventually paves the way to the ACES technologies, where the vehicles are expected to be shared while being autonomous,

FIGURE 2.2   Software-defined vehicles as the future of automotive.

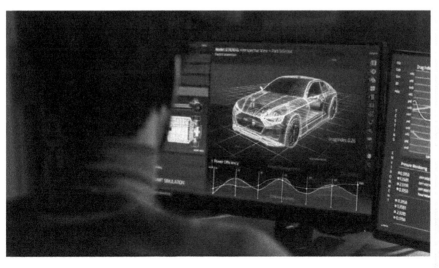

Gorodenkoff/Shutterstock.com.

connected, and electric—with software being the cornerstone of the technology.

2.4. Fourth Industrial Revolution

The Fourth Industrial Revolution (FIR), also known as Industry 4.0, has been concisely introduced in Chapter 1. However, it is worth citing again the FIR as a topic in this chapter as it is one of the most compelling factors in the evolution of ACES. As a matter of a fact, in the widely known discussion on the FIR by the World Economic Forum, one of the technologies mentioned is AVs [11, 12]. Despite the hypes and misconceptions that were reported on the AV production timeline, vehicle automation undeniably has a positive future [13]. With the advances in robotics boosting the FIR, we see a lot of investments made to support this shift. The FIR will disrupt the human way of life in many ways. With the Internet of Things (IoT) simultaneously progressing in the FIR, CVs development also facilitates and is paving way for shared mobility and sharing economy. This in return gives birth to increased discussions on shared mobility and reduced private vehicle ownership. To summarize, it is evident that the FIR has played a significant role in ACES growth.

2.5. United Nations Sustainable Development Goals

In 2015, the United Nations (UN) through one of its principal organs, the UN General Assembly, established the Sustainable Development Goals (SDGs). The aim is to delineate the needed shifts to produce a sustainable future for humankind, which inevitably has also been linked with the FIR topics [14].

The establishment of the SDGs, with other factors such as the FIR, is actively influencing and reshaping different industrial sectors. For automotive, namely, Elements 11 (Sustainable Cities and Communities) and 13 (Climate Action) of SDGs concomitantly motivate multidisciplinary researchers to delve deeper into the

ACES topics. As awareness of climate change concerns increases all over the world, it is not a coincidence the UN SDGs are also strengthening the measures of moving toward a more circular and sustainable economy. For example, with the UN SDGs 11, the topic "reducing pollution for a sustainable city" immediately suggests the need for vehicle electrification to reduce the carbon emissions produced by conventional vehicles [15].

Thus, it is not unforeseen when the market values with regard to UN SDGs are said to open up business opportunities with values at 12 trillion USD by 2030 [16]. UN SDGs not only promote the circular, green, and sustainable economy—which are also the trends that boost the ACES Vehicles development as Future Mobility—but also guide how the ACES as the Future Mobility could be developed as part of the larger concept of Smart City.

2.6. Society 5.0

Society 5.0 is a continuation of the FIR and SDGs discussions. But instead of focusing on the technology itself, the question raised is more on the "aftermath" of the FIR, i.e., discussing the society that will be living in the world when the said vision becomes reality. The conversations on Society 5.0 started in Japan where the Japanese government is heavily involved in the discussions [17]. Despite not mentioning ACES in specific, however, the discussions listed the need to create a caring human society that will live side by side with the emerging disruptive technologies [18, 19]. Therefore, the author of this book believes that Society 5.0 dialogues will eventually intersect with all other factors, thus facilitating a more ethical ACES technology productization and development.

2.7. Climate Change and Intergovernmental Panel on Climate Change 2021 Report

Climate change is one of the most popular recent scientific discussions. It is discussed not only by researchers and scientists, but it also influenced school pupils globally [20]. Among the examples of climate change topics include global warming, which resulted in the global change of temperature. Climate change is also one of the contributing causalities to the increasing number of bushfire incidents, with one of the latest examples happening in Australia and the United States of America (US) [21]. One of the contributors to climate change is industrial activities. The emissions of carbon from vehicles are also reported to be contributing to climate change. Thus, referring to Subsections 2.1–2.6, there is a necessity to formulate better and eco-friendly industrial activities [22] (Figure 2.3).

In 2021, the Intergovernmental Panel on Climate Change (IPCC) 2021 have published "Climate Change 2021: The Physical Science Basis," the first part of its IPCC Sixth Assessment Report [23, 24]. The report discussed climate change in detail, as well as the possible methods to mitigate it. The IPCC Sixth Assessment Report is the most up-to-date discussion on climate change, which is now amplifying the demand for transformation related to the visions of SDGs [25]. The aforementioned report was

FIGURE 2.3 Awareness of climate change has stimulated the ACES growth.

Roschetzky Photography/Shutterstock.com.

published when this book is now in the publication and preparation stage. Thus, there are still limited citations to analyze the impact of the reports on the automotive industry. However, it is highly evident that it will lead to more discussions on EVs, particularly the transition from Internal Combustion Engine (ICE) to vehicle electrification as it directly points to sustainability for a better climate [26]. This in return will strongly influence the ACES progress.

2.8. Carbon-Neutral and Finite Petroleum Resources

Compared to the previous industrial revolutions, the FIR has incorporated and is heavily laden with awareness of the societal and environmental effects. Studies have shown the importance of managing the effect of carbon emissions related to climate change (see Subsection 2.7). This has put pressure on businesses and companies as regards the topic. Consequently, companies and public organizations kickstarted carbon-neutral campaigns to strengthen the message and public awareness on this concern. For example, in 2018, Helsinki Airport officially achieved carbon-neutral status [27].

For the automotive industry, the advancements above have prompted further discussions on the future of ICEs. This has thereon indirectly boosted the market acceptance of electric vehicles. With the finite petroleum resources, we have even seen efforts by petroleum companies to diversify their business with renewable energies [28, 29].

Apart from that, not only in developed countries but also in developing states such as Indonesia, there are plans to lessen carbon emissions by introducing initiatives and plans to encourage the adoption of EVs [4]. Therefore, the author of this book believes that the demand for a sustainable economy, as well as finite petroleum resources, will trigger more discussions on ACES in the future and attract more investments.

2.9. Green, Circular, and Sustainable Economy

The concerns mentioned regarding current industrial activities toward the environment and nature have sparked the discussions of new economic concepts such as green and circular economy, which aim for sustainability. According to the European Parliament website, the circular economy is a model of production and consumption, which involves sharing, leasing, reusing, repairing, refurbishing, and recycling existing materials and products as long as possible [30]. With this philosophy, the circular economy has the potential to challenge climate change as well as other environmental pollution topics, among many others. Green economy on the other hand refers to an economy that is low carbon, resource efficient, and socially inclusive, according to the UN Environment Programme [31]. The economic size of these two is expected to be at least 12 trillion USD by 2030 [32]. Recently, a lot of start-ups are gaining traction and publicity for their efforts in promoting a circular and sustainable economy. For example, a start-up in Finland called Rens is trying to produce shoes from recycled coffee waste and plastic [33]. The concept might not be appealing if it was produced earlier; however, with the rising concerns and trends for sustainability, we are now seeing a growing number of similar businesses. For the automotive world, these developments have been subsequently encouraging the expansion of ACES, particularly with regard to EVs as it is expected to be the output of sustainable and circular automotive products.

2.10. Sharing Economy

According to the World Economic Forum, sharing economy means the sharing of underutilized assets, monetized or not, in ways that improve efficiency, sustainability, and community [34]. With the advances in Internet technology, the FIR, and digital awareness among world inhabitants, sharing economy has been a famous discussion for "future economy." Perhaps the most popular example, Airbnb Inc., has managed to indirectly disrupt the tourism ecosystem and hotel industry by allowing private landlords to lease their apartments to tourists. By cutting the middleman, Airbnb has facilitated the growth of sharing economy [34]. Similar to that, as mentioned, mobility platforms such as Uber, Gojek, Grab, and Bolt (previously Taxify) have boosted sharing economy applications in different regions [35, 36]. This has then supported the growth of shared mobility as part of ACES.

2.11. Regulations Facilitating ACES

ACES as a whole future mobility concept is an emerging technology, the discussions on its standardization are also ongoing. With the advances in vehicular automation and other topics mentioned above, a lot of standardization efforts have been carried out [37]. For example, SAE J3016 explicitly suggested the level of driving automation. It has now turned out to be one of the cornerstones for AV development by companies globally [38]. Some companies are even named in connection with the said levels of driving automation [39, 40]. In addition, as ACES technology is expected to output a lot of vehicle data, the privacy and cybersecurity concerns relating to data are heavily discussed as well [41]. Besides AVs, the discussions of regulations for vehicle connectivity are also discussed. For example, the European Union Agency for Cybersecurity (ENISA) has published recommendations for the security of Connected and Automated Mobility (CAM). The author of this book is grateful to be one of the contributors to the mentioned document [42]. These developments with regard to the new standardization for ACES have amplified the notion that ACES will be the future of mobility. More details on this will be discussed in the following chapters.

2.12. IoT and 5G

The term Internet and its origin can be traced even before the public audience widely used it. According to Merriam-Webster online dictionary, Internet means "an electronic communications network that connects computer networks and organizational computer facilities around the world" [43].

The widespread use of the Internet by the mass audience has only happened in recent decades. Nowadays, even older generations in developing countries are using smartphones [44]. Before 2010, most countries were still using slower Internet connections (e.g., dial-up Internet access). Nowadays, even for mobile phones, the median global Internet download speed performance is 24.26 Mbps (June 2021) [45]. This has allowed downloads of large-sized files to be done in a few seconds.

The advances of the Internet have then prompted the IoT explosions in society. This has permitted the interconnection between different devices with the support of cloud computing technology [46]. IoT has also reached automotive and mobility applications.

For example, the Internet of Vehicle (IoV) is the output of better vehicle connectivity. Consequently, data can be transferred between different vehicles. In addition, Vehicle-to-Everything (V2X), which allows vehicles to be connected to infrastructure, has also reached growth with the advances in the Internet and Connectivity Field [8].

For ACES to be safely deployed, vehicle navigation needs to be monitored during the AD trip. Thus, this requires reliable technology. 5G connection is reported to be one of the potential solutions for this [47]. In recent years, companies like Nokia, Ericsson, and Huawei, among many others, have collaborated with a lot of ACES companies to demonstrate the importance of 5G for future mobility purposes [48]. The growth of the telecommunication and connectivity fields, spearheaded by 5G and IoT, is believed to expedite the growth of ACES in the future.

2.13. Advances in Computing Power and Platforms

One of the crucial backbones of vehicle automation is the advancement in robotics algorithms research and development (R&D). When discussing removing human drivers from the vehicle for

AV, robust and safe algorithms are needed to replace human driving in the context of vehicle control. This algorithm should be at least compatible with the human driving quality, if not better. For example, when dealing with uncertainties in traffic, the algorithms should consider the risk of the surrounding conditions such as pedestrians, cyclists, and occluded objects. This requires the combination of comprehensive and sustainable object detection and motion planning and control algorithms, among many others [49].

However, the limitation of computing devices has always been the obstacle to more vehicle automation productization. Since the early 2010s, the advancements in computing platforms have enabled more R&D in ACES technology. Gradually over time, the economies of scale have allowed the platform to be cheaper, which in return will allow the democratization of the technology. Consequently, a lot of start-ups have been launched by researchers globally, developing their AV stack without the need for a spectacular budget to build their prototype [50]. Automotive original equipment manufacturers (OEMs) and carmakers such as TTTech Auto, NVIDIA, and Intel, among many others, are working to build high-end platforms for AD system computing [51].

In recent years, the advances in the computing medium have furthermore alleviated the prototype building of more advanced computing devices and now have advanced into the topics of quantum computing platforms (Figure 2.4). With quantum computing, it is anticipated that more complicated algorithms can be adopted for AD systems, which will increase the quality of future ACES technologies [52, 53]. Summarizing this section, the growth in this area is undeniable and will boost further ACES development.

2.14. Sensor Advancements (LiDAR, RADAR, Camera)

Autonomous technologies will see the removal of local human operators inside the vehicle during the AV operation. This suggests

FIGURE 2.4 Advancements in computing power and platforms helped in the development of complex automotive emerging technologies.

Arthur Palmer/Shutterstock.com.

the absence of not only human control of the vehicle but also human perceptions and instinct to measure the uncertainty on the road. This mandates a very reliable perception system of exogenous sensors. Furthermore, the AV technology will not be able to be productized if it only has strong simulation-based validated control systems but unreliable detection. Fortunately, the industries are now seeing rapid growth in automotive sensor industries. Sensors such as Light Detection and Ranging (LiDAR), Radio Detection and Ranging (RADAR), and camera have been actively commercialized to support the visions of ACES. Companies like Baraja and Blickfeld are among start-ups developing perception solutions for ACES technology, joining forces with other established companies like Velodyne and Qualcomm [54]. Considering the importance of perception for ACES technology, the growth of the industry is directly prompting more ACES progress.

2.15. Cross-Functional Software Product Development and Change Management

Technology is of limited benefit if it remains merely in laboratories and scientific publications. Cross-functional development is needed to bring out the whole system of ACES. For example, building a reliable collision avoidance system for an AV is not only the effort of the perception engineers, but instead, it also demands effort from the whole organization, including, to some extent, the legal department (e.g., where discussions about insurance for AD are needed) [55]. Furthermore, complex software development brings a different type of challenge compared to traditional automotive engineering. Thus, it is not a diminutive trend when we are seeing more discussions being made by automotive organizations and entities about "change management," "agile," "software architecture," and "Automotive Software Process Improvement and Capability dEtermination (ASPICE)," among many other related process and framework for the software product development [56, 57]. This progress shows that the importance of having a good process for productizing the ACES is realized by the practitioners, which will eventually propel the development further.

2.16. Silicon Valley and Rise of Start-Ups

ACES is a disruptive technology. With that being said, not only the technologies which need to be produced but also the corresponding organizational changes are also required to bring the technology as products to the market. With large organizations, the changes usually take time because they have such a large legacy of processes. But with "Silicon Valley" deep technology companies or start-ups, it is more straightforward and swifter to make the decisions for transformations, thus expediting the go-to-market of the new technology. It can be firmly suggested that start-up economics is one of the factors stimulating ACES growth. As ACES consists of a lot of complex technologies, this is a good sign with different companies trying to solve different pain points in the whole ACES architecture [58, 59].

What is unique about the previously mentioned trends like the FIR is that it also breaks down the traditional economic supply chain and ecosystem, where usually the major corporations are the leading forces of innovation. Rather, we have seen a lot of start-ups emerge in the market providing solutions with disruptive effects. Recent examples are Uber, which started as a start-up. Eventually, it disrupted the mobility industry and catalyzed the sharing economy. The growth of Uber also managed to coin the term "uberization," which fittingly paints an early vision of the future economy notion for transportation, i.e., shared mobility [60].

The progress not only happened in the developed nations but instead also in other regions such as the Southeast Asian (ASEAN) market. Grab and Gojek started in Malaysia and Indonesia, respectively, and are now both unicorns with company valuations of more than a billion USD. For ACES, a lot of start-ups have been entering the market too and sometimes even leading the transformation. A unique case is nuTonomy. What was started as a Massachusetts Institute of Technology spin-off start-up company then evolved into several entities, where it is now known as Motional, a household name in AV technology [61]. All these consequences stressed that not only the technology is changing but also the economic and automotive landscape. With more ACES-related start-ups getting acquired and listed on the stock exchange, it shows the demands in the automotive segments for ACES.

2.17. X-as-a-Service

The digitalization and excellent Internet connections with other tools such as smartphones spur new types of business such as sharing economics and X-as-a-Service (XaaS). With XaaS, companies tend to make profits from customers by providing values and services [62]. Companies like Netflix, for example, are among the best examples of a service business. Possibly at the end of the 2000s, people were still renting videos to watch movies (DVDs) at home, but with Netflix, they are paying minimum subscription costs to receive a vast amount of entertainment access. From the business perspective, with the absence of the need for physical stores, they can gain a lot of profit from this type of business. XaaS has also reached other fields which include the automotive industry [63]. With ACES, it is predicted that more evolution of XaaS will happen in the said industry, thus propelling the growth of the technology further.

2.18. Acqui-Hiring and M&A

One of the challenges faced by disruptive technologies (e.g., ACES) with regard to their productization is the need for talent from different fields (Figure 2.5). For example, CVs require talents from not only an automotive engineering background but also telecommunication and network engineering. The technology also demands an understanding of regulations and homologation from various

FIGURE 2.5   The war for talents has been seen in the automotive industry in recent years.

Vitalii Vodolazskyi/Shutterstock.com.

areas. This requires cooperation among various organizations and expertise. And therefore it is not surprising that start-up consolidations by different large organizations are happening in the industry. Many large corporations are buying deep technology start-ups to consolidate them into their organizations to solve the abovementioned pain point in the production process. The acquiring happens in two types: the first one is acqui-hiring and the second one is mergers and acquisitions (M&A). Acqui-hiring refers to when a company buys a company (usually start-ups) to obtain the talents in the start-ups. This is understandable as the efforts to recruit talents are not easy. Furthermore, convincing them to join a company and retain the employees are harder. Therefore, buying a start-up company with many extremely talented employees is sensible. And this is one of the recent events that have stimulated actively the growth of ACES development [64–66].

2.19. Supply Chain Evolutions

ACES will not exclusively deliver new technologies, but most importantly it is disrupting the economic chains of multiple industries. For example, as has been said, as the traditional OEMs are focusing on the hardware itself, most of the expertise for ACES comes from software-related companies, where the automotive software for future mobility is a complex field. The birth of unicorns related to ACES highlights the fact that ACES is here to stay, and this progress cannot be underestimated. The listings of EV companies such as NIO, Tesla, and XPeng in the major stock exchange also highlight this importance [67].

2.20. Digital Natives

The world is not only transforming technologically and economically, but in fact, the world will also see a shift in its inhabitants. In the year 2040, most of the people living and working can arguably be born after the year 2000. This new generation is called Digital Natives, which have been familiarized with the digital world since their childhood or even birth. With this stated, it is not a wonder that more and more technology has introduced digitalization to attract this kind of market audience. This has then encouraged more discussions of ACES as future mobility [68, 69].

2.21. Summary

This chapter has detailed twenty current trends and advances that are believed to propel the ACES growth further. As can be seen, it is not only technical progress, but instead, it is multi-field advancements that, if combined, yield a large-scale avalanche to boost the growth of ACES. In addition to the listed progress, other factors such as the ongoing global pandemic, the Coronavirus disease of 2019 (COVID-19), are also expected to influence the future mobility industry direction. The author believes that this list will paint a suggestive picture, especially to the general reader, of the scale of the disruptions that will be faced by the automotive industry. In the next chapters, detailed specifications for each ACES element are written, and by reading this chapter, the readers will have an overview of the current automotive industry prior to that.

References

1. Fleming, W., "Forty-Year Review of Automotive Electronics: A Unique Source of Historical Information on Automotive Electronics," *IEEE Vehicular Technology Magazine* 10, no. 3 (2015): 80-90.

2. Hamid, U.Z.A., Zakuan, F.R.A., Zulkepli, K.A., Azmi, M.Z. et al., "Autonomous Emergency Braking System with Potential Field Risk Assessment for Frontal Collision Mitigation," in *2017 IEEE Conference on Systems, Process and Control (ICSPC)*, Melaka, Malaysia, 71-76, IEEE, December 2017.

3. European Parliament, "Parliament Approves EU Rules Requiring Life-Saving Technologies in Vehicles," accessed August 2022, https://www.europarl.europa.eu/news/en/press-room/20190410IPR37528/parliament-approves-eu-rules-requiring-life-saving-technologies-in-vehicles.

4. Hamid, U.Z.A., Ishak, S.Z., and Imaduddin, F., "Current Landscape of the Automotive Field in the ASEAN Region: Case Study of Singapore, Malaysia and Indonesia—A Brief Overview," *Asean Journal of Automotive Technology* 1, no. 1 (2019): 21-28.

5. Razak, S.F.A., Yogarayan, S., Azman, A., Abdullah, M.F.A. et al., "Driver Perceptions of Advanced Driver Assistance Systems: A Case Study," *F1000Research* 10, no. 1122 (2021): 1122.

6. Kusano, K.D. and Gabler, H.C., "Safety Benefits of Forward Collision Warning, Brake Assist, and Autonomous Braking Systems in Rear-End Collisions," *IEEE Transactions on Intelligent Transportation Systems* 13, no. 4 (2012): 1546-1555.

7. Keall, M.D., Fildes, B., and Newstead, S., "Real-World Evaluation of the Effectiveness of Reversing Camera and Parking Sensor Technologies in Preventing Backover Pedestrian Injuries," *Accident Analysis & Prevention* 99 (2017): 39-43.

8. Hamid, U.Z. Abdul, H.Z., and Limbu, D.K., "Internet of Vehicle (IoV) Applications in Expediting the Implementation of Smart Highway of Autonomous Vehicle: A Survey," in Al-Turjman, F. (ed.) *Performability in Internet of Things* (Cham: Springer, 2019), 137-157.

9. Vailshery, L.S., "Size of the Automotive Software Market Worldwide from 2018 to 2025," Statista, accessed August 2022, https://www.statista.com/statistics/590055/worldwide-automotive-software-market-size/.

10. Deloitte, " Software-Defined Vehicles—A Forthcoming Industrial Evolution," accessed August 2022, https://www2.deloitte.com/cn/en/pages/consumer-business/articles/software-defined-cars-industrial-revolution-on-the-arrow.html.

11. Schwab, K., "The Fourth Industrial Revolution: What It Means, How to Respond," World Economic Forum, accessed August 2022, https://www.weforum.org/agenda/2016/01/the-fourth-industrial-revolution-what-it-means-and-how-to-respond.

12. Schwab, K., *The Fourth Industrial Revolution* (New York: Currency Books, 2017)

13. Kaivo-oja, J., Lauraeus, T., and Knudsen, M.S., "Picking the ICT Technology Winners-Longitudinal Analysis of 21st Century Technologies Based on the Gartner Hype Cycle 2008-2017: Trends, Tendencies, and Weak Signals," *International Journal of Web Engineering and Technology* 15, no. 3 (2020): 216-264.

14. Pedersen, C.S., "The UN Sustainable Development Goals (SDGs) Are a Great Gift to Business!" *Procedia CIRP* 69 (2018): 21-24.

15. Wolff, S., Brönner, M., Held, M., and Lienkamp, M., "Transforming Automotive Companies into Sustainability Leaders: A Concept for Managing Current Challenges," *Journal of Cleaner Production* 276 (2020): 124179.

16. United Nations, "UN Secretary-General's Strategy for Financing the 2030 Agenda," accessed August 2022, https://www.un.org/sustainabledevelopment/sg-finance-strategy.

17. Cabinet Office, Government of Japan, "Society 5.0," accessed August 2022, https://www8.cao.go.jp/cstp/english/society5_0/index.html.

18. Potočan, V., Mulej, M., and Nedelko, Z., "Society 5.0: Balancing of Industry 4.0, Economic Advancement and Social Problems," *Kybernetes* 50, no. 3 (2021): 794-811.

19. Deguchi, A., Hirai, C., Matsuoka, H., Nakano, T. et al., "What Is Society 5.0," *Society* 5 (2020): 1-23.

20. Aiyedun, T.G., "Effect of Animation Teaching Strategy on Secondary School Students' Achievement, Retention and Interest in Climate Change in Lokoja, Kogi State," *International Journal of Trend in Scientific Research and Development (IJTSRD)* 4, no. 3 (2020): 944-949.

21. Smith, A.J.P., Jones, M.W., Abatzoglou, J.T., Canadell, J.G. et al., "Climate Change Increases the Risk of Wildfires: September 2020," ScienceBrief, 2020.

22. Jones, E.C. and Leibowicz, B.D., "Contributions of Shared Autonomous Vehicles to Climate Change Mitigation," *Transportation Research Part D: Transport and Environment* 72 (2019): 279-298.

23. IPCC, "Climate Change 2021: The Physical Science Basis," accessed August 2022, https://www.ipcc.ch/report/sixth-assessment-report-working-group-i/.

24. IPCC, "Sixth Assessment Report," accessed August 2022, https://www.ipcc.ch/assessment-report/ar6/.

25. IPCC, "Climate Change 2021: The Physical Science Basis," in Masson-Delmotte, V., Zhai, P., Pirani, A., and Connors, S.L. et al. (eds.), *Contribution of Working Group I to the Sixth Assessment Report of the Intergovernmental Panel on Climate Change* (Cambridge: Cambridge University Press, 2021).

26. Li, C., Cao, Y., Zhang, M., Wang, J. et al., "Hidden Benefits of Electric Vehicles for Addressing Climate Change," *Scientific Reports* 5, no. 1 (2015): 1-4.

27. Finavia Newsroom, "Helsinki Airport Is Carbon Neutral. Here Are 5 Concrete Actions That Ensure Carbon Neutrality," accessed August 2022, https://www.finavia.fi/en/newsroom/2018/helsinki-airport-carbon-neutral-here-are-5-concrete-actions-ensure-carbon-neutrality.

28. Shell, "Renewables and Energy Solutions," accessed August 2022, https://www.shell.com/energy-and-innovation/new-energies.html.

29. Hartmann, J., Inkpen, A.C., and Ramaswamy, K., "Different Shades of Green: Global Oil and Gas Companies and Renewable Energy," *Journal of International Business Studies* 52, no. 5 (2021): 879-903.

30. European Parliament News, "Circular Economy: Definition, Importance and Benefits," accessed August 2022, https://www.europarl.europa.eu/news/en/headlines/economy/20151201STO05603/circular-economy-definition-importance-and-benefits.

31. UN Environment Programme, "Green Economy," accessed August 2022, https://www.unep.org/regions/asia-and-pacific/regional-initiatives/supporting-resource-efficiency/green-economy.

32. Hoek, M., *The Trillion Dollar Shift* (London: Routledge, 2018)

33. Rens Original, "World's 1st Coffee Sneaker," accessed August 2022, https://rensoriginal.com/.

34. Rinne, A., "What Exactly Is the Sharing Economy," World Economic Forum, vol. 13, 2017, accessed August 2022, https://www.weforum.org/agenda/2017/12/when-is-sharing-not-really-sharing/.

35. Zervas, G., Proserpio, D., and Byers, J.W., "The Rise of the Sharing Economy: Estimating the Impact of Airbnb on the Hotel Industry," *Journal of Marketing Research* 54, no. 5 (2017): 687-705.

36. Kurniawati, D.E. and Khoirina, R.Z., "Online-Based Transportation Business Competition Model of Gojek and Grab," *Advances in Social Science, Education and Humanities Research* 436 (2020): 1054-1057.

37. Danks, D. and London, A.J., "Regulating Autonomous Systems: Beyond Standards," *IEEE Intelligent Systems* 32, no. 1 (2017): 88-91.

38. Shuttleworth, J., "SAE Standards News: J3016 Automated-Driving Graphic Update," SAE International, accessed August 2022, https://www.sae.org/news/2019/01/sae-updates-j3016-automated-driving-graphic.

39. Sensible 4 Website, "A Sensible Approach to Self-Driving," accessed August 2022, https://sensible4.fi.

40. Five's Website, "Faster Development, Quicker Coverage, Better Analysis," accessed August 2022, https://www.five.ai.

41. Costantini, F., Thomopoulos, N., Steibel, F., Curl, A. et al., "Autonomous Vehicles in a GDPR Era: An International Comparison," in Milakis, D., Thomopoulos, N., and van Wee, B. (eds.), *Advances in Transport Policy and Planning*, vol. 5 (Cambridge, MA: Academic Press, 2020), 191-213.

42. European Union Agency for Cybersecurity, "Recommendations for the Security of CAM," ENISA, accessed August 2022, https://www.enisa.europa.eu/publications/recommendations-for-the-security-of-cam.

43. Merriam-Webster Dictionary, "Internet," accessed August 2022, https://www.merriam-webster.com/dictionary/Internet.

44. Glushkova, S., Belotserkovich, D., Morgunova, N., and Yuzhakova, Y., "The Role of Smartphones and the Internet in Developing Countries," *Revista ESPACIOS* 40, no. 27 (2019): 10.

45. Speedtest's Website, "Speedtest Global Index," accessed August 2022, https://www.speedtest.net/global-index.

46. Nord, J.H., Koohang, A., and Paliszkiewicz, J., "The Internet of Things: Review and Theoretical Framework," *Expert Systems with Applications* 133 (2019): 97-108.

47. Kotilainen, I., Händel, C., Hamid, U.Z.A., Santamala, H. et al., "Arctic Challenge Project's Final Report: Road Transport Automation in Snowy and Icy Conditions," Väyläviraston tutkimuksia, 2019.

48. Peters, M.A. and Besley, T., "5G Transformational Advanced Wireless Futures," *Educational Philosophy and Theory* 53, no. 9 (2021): 847-851.

49. Hamid, U.Z., Abdul, Y.S., Zamzuri, H., Rahman, M.A.A. et al., "A Review on Threat Assessment, Path Planning and Path Tracking Strategies for Collision Avoidance Systems of Autonomous Vehicles," *International Journal of Vehicle Autonomous Systems* 14, no. 2 (2018): 134-169.

50. Theisen, N., "Continental's Venture into Corporate Venture Capital: How Startup Investments May Help a Large Automative Incumbent to Deal with Disruptive Transformations in the Auto Space," PhD dissertation, Nova School of Business and Economics, 2019.

51. Suk, J.H. and Lyuh, C.G., "State-of-the-Art AI Computing Hardware Platform for Autonomous Vehicles," *Electronics and Telecommunications Trends* 33, no. 6 (2018): 107-117.

52. Kashyap, K., Shah, D., and Gautam, L. "From Classical to Quantum: A Review of Recent Progress in Reinforcement Learning," in *2021 2nd International Conference for Emerging Technology (INCET)*, Belgaum, India, 1-5, IEEE, 2021.

53. Gardiner, M., Truskovsky, A., Neville-Neil, G., and Mashatan, A., "Quantum-Safe Trust for Vehicles: The Race Is Already On," *Queue* 19, no. 2 (2021): 93-115.

54. Tsai, D., Worrall, S., Shan, M., Lohr, A. et al., "Optimising the Selection of Samples for Robust Lidar Camera Calibration," arXiv preprint arXiv:2103.12287, 2021.

55. Dhar, V., "Equity, Safety, and Privacy in the Autonomous Vehicle Era," *Computer* 49, no. 11 (2016): 80-83.

56. Hamid, U.Z.A., Mehndiratta, M., and Adali, E., "Adopting Aviation Safety Knowledge into the Discussions of Safe Implementation of Connected and Autonomous Road Vehicles," SAE Technical Paper 2021-01-0074, 2021, https://doi.org/10.4271/2021-01-0074.

57. Todorovic, M., Simic, M., and Kumar, A., "Managing Transition to Electrical and Autonomous Vehicles," *Procedia Computer Science* 112 (2017): 2335-2344.

58. Bezruchonak, A., "Driverless Mobility and the Geographic Analysis of Contemporary Autonomous Vehicles Startup Ecosystem," *Prace Komisji Geografii Komunikacji PTG* 21, no. 4 (2018): 7-13.

59. Liu, S., "Critical Business Decision Making for Technology Startups: A Percept in Case Study," *IEEE Engineering Management Review* 48, no. 4 (2020): 32-36.

60. Rahman, A., Airini, U.Z.A.H., and Chin, T.A., "Emerging Technologies with Disruptive Effects: A Review," *Perintis eJournal* 7, no. 2 (2017): 111-128.

61. Cusumano, M.A., "Self-Driving Vehicle Technology: Progress and Promises," *Communications of the ACM* 63, no. 10 (2020): 20-22.

62. Duan, Y., Fu, G., Zhou, N., Sun, X. et al., "Everything as a Service (XaaS) on the Cloud: Origins, Current and Future Trends," in *2015 IEEE 8th International Conference on Cloud Computing*, New York City, NY, 621-628, IEEE, 2015.

63. Goodall, W., Dovey, T., Bornstein, J., and Bonthron, B., "The Rise of Mobility as a Service," *Deloitte Rev* 20 (2017): 112-129.

64. Polsky, G.D. and Coyle, J.F., "Acqui-Hiring," *Duke Law Journal* 63, no. 2 (2013): 281-346.

65. Chui, M., "Artificial Intelligence the Next Digital Frontier," McKinsey and Company Global Institute, vol. 47, no. 3.6, 2017.

66. Delke, V.F., "Identification of Startups as Innovation Partners: Analyzing Complex Search Strategies within the Automotive Industry," Master's thesis, University of Twente, 2017.

67. Fartaj, S.-R., Kabir, G., Eghujovbo, V., Ali, S.M. et al., "Modeling Transportation Disruptions in the Supply Chain of Automotive Parts Manufacturing Company," *International Journal of Production Economics* 222 (2020): 107511.

68. Tran, T., Ho, M.-T., Pham, T.-H., Nguyen, M.-H. et al., "How Digital Natives Learn and Thrive in the Digital Age: Evidence from an Emerging Economy," *Sustainability* 12, no. 9 (2020): 3819.

69. Koehler, C., Appel, D., and Beck, H., "Winning Strategies in the Race for Connected Autonomous Cars," *Auto Tech Review* 6, no. 1 (2017): 36-41.

SECTION 3
Concise Overview of ACES

Autonomous Vehicles: Concise Overview

3.1. Background and What

Fiction turns into reality. Art navigates the future. We have seen a lot of quotes and remarks similar to these. And we are seeing a lot of elements of the development of new technology that are highly similar to those shown in science fiction arts and films. Is it just a coincidence? One might question. Maybe yes, maybe not [1, 2]. Nonetheless, when discussing future emerging technologies, what frequently appears in the minds of the general public is a glimpse of a futuristic utopic vision, highly influenced by science fiction movies [3]. Of course, this is a misconception. Therefore, what exactly is "autonomous driving"? Even in the organizations and industries that are working on AV technology, the distinction between different terms describing the technology which enables mobility without a human driver is vague and ambiguous [4, 5]. On some occasions, underestimating the importance of interpreting these definitions can lead to accidents because of unclear and divided perspectives throughout the technology development. Therefore, before moving on to more detailed discussions on AVs, it is good to briefly highlight the difference between the various definitions and terms for this topic. The table below illustrates the summary, which has been reworded and paraphrased from Hamid et al. [6], Park et al. [7], and the SAE J3016 Visual Chart "Levels of Driving Automation" [8].

For more details on the discussions in <u>Table 3.1</u>, the author strongly suggests a further reading of the article by Hamid et al. [<u>6</u>] and other references cited in the mentioned articles. From the discussions in the table and references above, the AV is a mobility platform with the ability to move from point A to point B, depending on the level of automation, without the need for human intervention. The level of automation for road vehicles can be defined by several metrics, the most prominent being the SAE Levels of Driving Automation™, which is based on the SAE J3016™ Recommended Practice: Taxonomy and Definitions for Terms Related to Driving Automation Systems for On-Road Motor Vehicles [<u>9</u>]. It is important to note that for this book, the author might use the terms interchangeably. The readers can always look to <u>Table 3.1</u> as the point of reference. Furthermore, please note that the AV technology mentioned in this book are specifically for road vehicle use case and not for other applications such as Autonomous Mobile Robots (AMR), maritime autonomous surface vessels, and Vertical Take-Off and Landing (VTOL) aircraft.

As the cornerstone of the AV technology is the AD software, in the next section, the author provides the definition and background of each component in the software stack. The author

TABLE 3.1 Definitions of different terminology related to the AV topic [<u>6</u>–<u>8</u>].

Phrase	Denotation
Automated Vehicle	A vehicle that can operate without human input, at least in hazardous scenarios.
Self-Driving Vehicle	A vehicle that can operate without human intervention. Human operators are still monitoring and can intervene in the process if needed.
Autonomous Vehicle	A vehicle that functions, perceives its surroundings, and moves without the need for human involvement. In edge cases and hazardous scenarios, the AV should be able to provide a collision avoidance maneuver on its own without the need for human assistance.
Driverless Vehicle	This terminology has the closest synonym to an AV, but with a more explicit expression. A driverless vehicle should be able to fully operate with its intelligence in all circumstances. Therefore, no human monitoring and intervention is expected.

© SAE International.

believes it will help the reader, especially those who are of non-technical backgrounds, to understand better the back-end nature of AD technology. The overview will briefly provide a suggestive concept rather than being exhaustive.

3.2. Technical Overview

There are a lot of rationales around the motivation of mankind to automate more of the aspects of their daily lives. According to Chui et al. [10], automation will not only bring savings in labor costs but also improve the skills of the human workforce, allowing the focus to be utilized and redirected into topics of higher value. This eventually brings a lot of financial and societal benefits. Furthermore, because automation can reduce the tasks and energy consumed by humans, this will also improve the quality of life of employees. Yet unreliable industrialization of automation technology can cause capital investment because of safety factors, among many others. That is why, in the context of the productization of autonomous systems, knowledge of the software back end needs to be emphasized as well. This is to enable reliable automation technologies (including the AD) that can achieve its initial objective, i.e., to facilitate a better life for mankind.

The AV software stack incorporates several components, including "mapping and localization," "environmental awareness," "risk assessment," "motion planning," and "motion control." In addition to these modules, there are also "peripheral" and "auxiliary" modules such as "remote monitoring" as well as "user interface" that have been developed by several entities [11, 12].

3.2.1. How Humans Operate: An Anecdote to Simplify the AV Back-End Algorithms

To better comprehend the operation of an AV from a layman's non-technical perspective, let us take a look at this oversimplified anecdote on how a human operates. The author reasons that including this anecdote in this chapter will allow the non-technical background readers to better understand this topic.

Let us consider that a male person is in an empty room. Knowledge of the room space is comparable to the mapping component of an autonomous system. This information is stored earlier in his brain. His current position in the room is further assessed using the perception information (eyes) and translated as his actual coordinate in the room by his brain. When he is trying to move from point A to point B in the room, his brain plans the needed trajectory and computes the desired motion, which will eventually be actuated by his actuators, which are his feet. During the maneuver, if he sees some unknown objects on his path, he replans his trajectory with the input from his eyes (perception), therefore performing the desired obstacle avoidance behavior with his feet. Figure 3.1 below illustrates this anecdote.

3.2.2. Human-Operated Vehicles Operation

Following the discussions above, let us take a look at the operation of a human-driven vehicle. For the scenario given in Figure 3.2, a vehicle is driven by a human on a straight road where suddenly a

FIGURE 3.1   Oversimplification of how a human operates as a metaphor compared to autonomous systems.

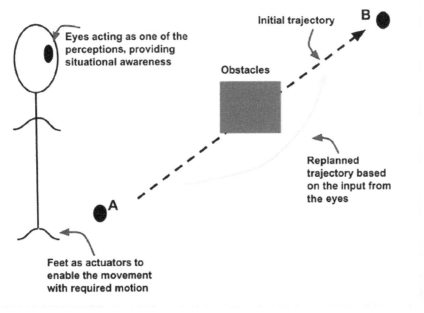

© SAE International.

FIGURE 3.2 Human-driven vehicle operation illustration in nominal driving condition and collision avoidance situation.

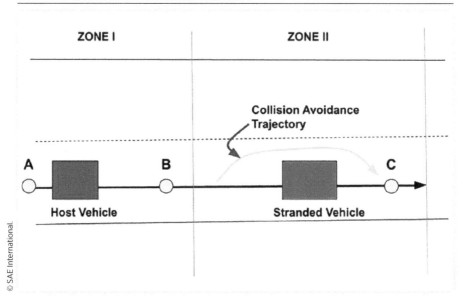

© SAE International.

previously occluded obstacle appears on the road. In Zone I, the vehicle traverses through points A to B, where the human driver operates the actuators of the vehicle, i.e., braking, gas, and steering wheel to control the longitudinal and lateral motions of the host vehicle. Meanwhile, the information about the surroundings and localization in its environment is perceived by the human. From point B to point C in Zone II, however, once the person notices the previously unknown object on the road lane, he replans his trajectory with the input from his perception (eyes) and subsequently performs the desired collision mitigation behavior with either braking, steering, or controlling the vehicle accelerations.

Comparing Figures 3.1 and 3.2, with AVs, the platforms still ought to accomplish the same behavior as mentioned in the previous subsections, as well as maintain the same safety requirements. However, the distinction is that for the AV, the human driver is being taken out of the equation. Thus, the AV needs to be able to operate with all the synthetic perception and intelligence, assuring safety and comfort at all times. How could this be done? To yield the desired AV behavior, all modules in the AD software stack should be developed properly, assuring a reliable, feasible, and secure AD experience [13]. In the next subsections, the author performs a deep dive into the core AV technologies. This will assist software engineers of non-robotics background to develop the system better by recognizing the impact of inefficient algorithms in the AD software. As a note, for brevity, this book only focuses on the high-level software of the AV, instead of other modules such as low-level control and vehicle hardware platform software.

3.2.3. AV Stack in General

In Figure 3.3 below [14], a generic architecture of an autonomous system is illustrated. The input to this system comes from a variety of sensors which include cameras, LiDAR, and RADAR, among many others. The merged data information is then utilized by several perception modules for different uses such as "mapping and localization" as well as "environmental awareness." The high-level control systems comprise the guidance and navigation modules, formulating the continuous vehicle behavior. These outputs will then be utilized by the low-level control of the vehicle, which eventually enables the vehicle to maneuver. This is of course an oversimplification of a really complex system, but it is beneficial to understand the big picture first before we are going to review each of the components in the next subsections.

FIGURE 3.3   Generic diagram of AD software architecture [14].

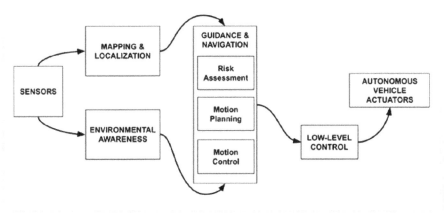

© SAE International.

3.2.4. Mapping and Localization

For an AV, it is extremely important that the vehicle knows its whereabouts in a specific location at a particular period in time. Failing to do this will render the vehicle unable to generate a safe and reliable AD behavior [15]. Compared to the indoor AMR, most of the time the AV encounters challenges because of the need to map and localize itself in outdoor areas. These include the challenging outdoor terrain, landmark unavailability or absence, and bad weather.

For the AV to be able to navigate itself reliably, it needs to be able to receive accurate information from its sensors (which include the sensor positions on the physical frame of the vehicle), the map of its environment, and its correct positioning information within the said map. All of this then will provide the AV with the required vehicle state information such as the coordinates as well as other motion states.

Figure 3.4 below simplifies the mapping and localization process sequence of an AV. Using the information from the exogenous

FIGURE 3.4 Oversimplified mapping and localization process sequence [16–23].

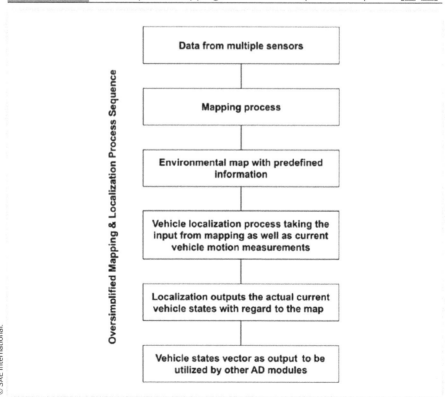

© SAE International.

perception sensors such as LiDAR, RADAR, and camera, data are then taken into a mapping process which incorporates algorithms to construct the environmental map. In the regular scenario, this map also holds predefined information about the environment such as landmarks and static infrastructures. Having a map alone is just the beginning. For an AV to operate, among the essential features are for it to know its whereabouts on the defined map. This ability is called positioning, sometimes also dubbed localization. To avoid confusion, let us take a look at the difference between the two terminologies. Positioning itself is the formulation of the AV coordinates and positions, whereas localization refers to the act of the AV to localize itself on a given map. The self-driving vehicle needs to be able to localize itself, where the output is the actual state of the vehicle, which include current vehicle coordinates and vehicle motion (e.g., multidirectional accelerations).

With the output from mapping and localization, the other modules such as planning and control can employ the information for the guidance, navigation, and control functions and operations. Therefore, as written above, mapping and localization can be perceived to be one of the extremely vital components of the AD stack. The module needs to always output reliable vehicle states, regardless of the environmental condition. As the main aim of this chapter is to provide an overview of the ACES technology, and not the discussions of the technical details, readers are suggested to read the following state-of-the-art surveys and articles on the said AD software component and its algorithms [16–23].

3.2.5. Environmental Awareness

While the "mapping and localization" answers the questions "Where is the vehicle now?," during the AD navigation, to compensate for the absence of human driving, the question of "What is surrounding the vehicle throughout the journey?" should also be answered. If the mapping process provides the predefined knowledge of the environment, and the localization module formulates the actual vehicle states and information in a certain instance, the environmental awareness system facilitates the driverless host vehicle with a vision of its surroundings. This means that during the AV navigation, unknown objects and changes in the map are monitored, detected, and tracked by the online perception modules of the environmental awareness systems. Environmental awareness usually consists of "detection" and "tracking" components [24–26].

Object and environment detection is required to perceive the objects and changes in the environment. With detection, the module confirms the existence of previously unknown objects. The object tracking module then formulates the tracked object states and information. For example, the surrounding object coordinates, its velocity, accelerations, and possible dimensions. Together with the mapping and localization system, environmental awareness is very important information to the subsequent AV behavioral control modules. As AV will be used for on-road applications, the highly dynamic road traffic environment should be included in the consideration of environmental awareness system development. Unreliable detection and tracking modules might cause fatalities because of the ineffectiveness of sensing potential hazardous entities on the road. Figure 3.5 illustrates the different types of objects and elements that should be monitored, detected, and tracked by the environmental awareness module, which includes

FIGURE 3.5 Different considerations and requirements for a reliable and safe environmental awareness system for AVs [27, 28].

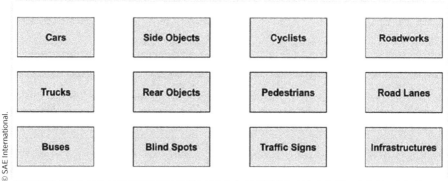

vulnerable road users (VRU) such as pedestrians and cyclists. As can be seen, the considerations do not only consist of the different objects, but also different traffic scenarios as well as blind-spot monitoring [27, 28].

By having the information and knowledge about the potentially hazardous object, the vehicle will be able to precisely plan the needed collision mitigation maneuver. It can be deduced that an environmental awareness module should be able to provide the AV with continuous, reliable, and robust functionalities in varied road traffic scenarios with different unknown and uncertain objects. For interested readers, more in-depth technical discussions for the environmental awareness modules of autonomous platforms can be found in these references [24, 29–32].

3.2.6. Risk Assessment

Up until the previous sections, we are still examining the perception of an AV (map, localization, and environmental awareness). With the output from the aforementioned system, the AV should then unassistedly decide its next behavior throughout the driverless journey. The switching between different behaviors requires an accurate and timely decision-making strategy. The complex decision-making takes the input from a module, called "risk assessment," which, based on the perception information, will provide safety metrics and thresholds for ensuing changes of maneuver. In this and the following subsections, we will discuss the intelligence of the AV, i.e., the brain of the AV. The first module that will be inspected is "risk assessment."

One of the blockers which avert the adoption of AV by the public is because of safety coniderations [33]. The AV, at all times and at all costs, should be able to maintain the required safety metrics for the potential hazard. In addition, it should be able to formulate the risk for multiple varied scenarios, which include the static, dynamic, and uncertain environment. According to Hamid et al. [34], static refers to an object that is not moving, like a parked car on the side of the road. Hazardous dynamic objects meanwhile require the collision avoidance maneuver with moving objects by

FIGURE 3.6 Risk assessment should be able to provide the risks of different origins, such as occluded objects, lane departure, and oncoming vehicles in the opposite lanes [35].

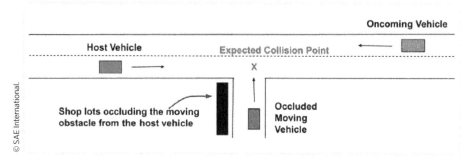

the AV. For example, the overtaking of a moving frontal object. Finally, the driverless vehicle should be able to avoid and mitigate the risks of the uncertain environment, such as the sudden appearance of previously occluded objects. In the work by Hamid et al. [35], Artificial Potential Field strategy is proposed and adopted into a collision avoidance system as a risk assessment method to measure the risk of suddenly appearing occluded objects from potential crash at intersections. As can be seen in Figure 3.6, in this particular scenario, risk assessment not only needs to be able to provide a safety threshold that will trigger the required emergency maneuver (either braking, steering, or both), but it also needs to make sure that the host vehicle is not deviating too far from its lane while doing so (in case of the risks of another oncoming vehicle in the opposite lane). This is done by measuring the risks of collisions, lane departure, and road environment. Therefore, a reliable risk assessment is crucial in aiding the motion planning and control in yielding a safe AD behavior, and this requires the measurement of different aspects of risks. Further discussions can be found in the work by Hamid et al. [35].

Assuming dependable mapping, localization, and environmental awareness modules are present, the risk assessment task is to formulate the safety metrics for the object (e.g., safe distance and safe time). To provide a better understanding to the general audience, the formulation and concept of threat assessment strategy are shown in Figure 3.7 [36].

FIGURE 3.7   Risk assessment keeps the safety metrics from the potential collision point [34, 36].

© SAE International.

From Figure 3.7, in each instance, the AV should be able to calculate the safety metrics to the potential collision point, which is done in the threat assessment module. In this example, the Time-to-X formulations are adopted. If the threshold is violated, the threat assessment will then trigger the decision-making in the behavior planning of the AV, which will then be the input for the trajectory replanning for the collision avoidance maneuver. Apart from Time-to-X, there are a lot of strategies that can be used for threat assessment purposes. Eventually, the main aim is to maintain the safety metrics to the potential collision point [37].

Furthermore, risk assessment not only ought to provide safety metrics, but a good and reliable risk assessment module will eventually sustain the comfort of the vehicle, especially in the context of longitudinal motion replanning during the adaptive cruise control behavior. The consistent comfort and safety feature of an AD ride will eventually increase user acceptance of the technology [38].

3.2.7. Motion Planning

The AV not only must be able to monitor its environment and position, track the surrounding object, and formulate required safety metrics for the potential risk, but instead, it also needs to calculate a reliable and safe trajectory in varied scenarios. This is done in the motion planning module. Motion Planning is the component responsible to formulate the needed replanned motion in the collision avoidance system. It is important to mention the difference between "path planning," "trajectory planning," and "motion planning." Therefore, the interested readers are advised to do further readings by Hamid et al. [34], Sugihara and Smith [39], González [40], Katrakazas [41].

For simplicity, this article opted to use the term motion planning for most of the time, i.e., planning of the vehicle motion as the input that will be tracked by motion control strategy. However, motion planning is not always about "replanning." In a controlled environment where no potential risks are detected by the environmental awareness and no threats are calculated by the risk assessment, an AV can navigate by merely following a predefined reference trajectory, for example, in the case of the automated guided vehicle doing routine-based tasks in a factory. However, for on-road AD, this is not feasible as the road traffic is extremely dynamic and uncertain. Therefore, in this subsection, the author will give focus on the "motion planning" application for a collision avoidance system for AVs.

In a situation that requires collision avoidance actuation, as shown in Figure 3.3 earlier, the output from "mapping and localization" and "environmental awareness" are utilized as the input to the risk assessment as well as motion planning. The information consists of the current state of the host vehicle and the surrounding obstacles states. Risk assessment then provides the motion planning with the additional safety metrics. These are then utilized by the motion planning to calculate a feasible, reliable, and safe replanned trajectory for the vehicle behavior.

- *Feasible* trajectory means that the replanned trajectory should consider the nonholonomic nature of most road vehicles, thus formulating a trajectory that is feasible to be traversed.

- *Reliable* means that the trajectory should be continuous and not discrete. This will help the AV to have a comfortable ride without unnecessary jerks in the motion.

- *Safe* means that the replanned trajectory should always consider the safety metrics provided by the risk assessment, thus assuring that there will be no additional risks during the collision avoidance maneuver. This newly replanned trajectory will then override the initial trajectory of the AV. Once the collision has been mitigated or avoided, and the safety metrics are more than the threshold, the behavior planning will dictate the motion control to follow the initial trajectory again.

Concluding this section, for a feasible, reliable, and safe motion to be planned, factors such as the different dynamics for different AV platforms and distinctive road surfaces should be considered. Finally, the aim of motion planning, particularly in the collision avoidance situation, is to output a dependable motion in the multi-traffic scenario for on-road AV purposes. Therefore, considering a lot of dependencies of the systems on other modules, this is not an easy job that should be taken lightly. More discussions on motion planning can be found in this comprehensive survey by Lefèvre et al. [42] and Claussmann et al. [43].

3.2.8. Motion Control

The planned motion (or in the case of collision avoidance scenario, the replanned motion), will then be published to the Motion Control module. In the module, the high-level trajectory tracking system then assures that the vehicle will follow the desired behavior while reducing the jerk and maintaining the comfort of the vehicle. As it is the final component in the AD-software pipeline, as can be seen in Figure 3.3, it is highly dependent on the previous modules.

Furthermore, because motion control largely concerns control and algorithms, therefore usually control engineers work with the assumption that the other modules perform well. To expedite the development timeline of the module and prevent the bottleneck because of the dependence on other modules, a lot of simulator-based verification and validation tools have been adopted to assist the motion control development, such as IPG CarMaker as well as CarSim by Mechanical Simulation Corporation [44].

Figure 3.8 shows the simplified schematic of motion control. The trajectory tracking functions by taking the desired motion (yaw rate, deceleration, and acceleration) and the current vehicle states as the input from the motion planning and the localization modules. Trajectory tracking then penalizes the error between the desired and tracked motion, eventually formulating the input to the low-level control (desired steering angle, braking, and accelera-tion torques). Low-level control then calculates the actual actuation

FIGURE 3.8   Brief diagram of high-level motion control for AVs [34, 36].

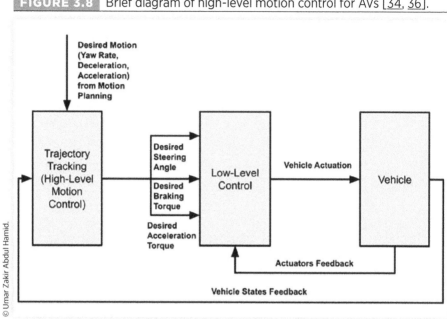

metrics for the AV actuators (steering, braking, and throttle) [34, 36].

For AVs, a lot of consideration needs to be done when formulating the motion control, for example, the weight, speed, dynamics, and physical dimensions of the platforms. Thus, a good control strategy must guarantee the feasibility of the vehicle maneuver by tracking the reference path unfailingly throughout the AD trip. For collision avoidance application, the usage of more robust and optimal control is encouraged to handle the nonlinearity caused by the sudden changes in multidirectional motion during the emergency maneuver. Further readings for the motion control of AVs can be found in the works by Wang et al. [45], Klomp et al. [46], and Iskander et al. [47].

3.2.9. Interface, Calibration, and Monitoring

As of now, we have discussed the core algorithms components of AV. However, for AD technology, there are other components and addended interfaces within the software that also needs to be materialized too. For example, in the human-machine interface, as the road traffic that will be utilized by the AVs is not only used by the driverless platforms alone, the AV should also be designed to be able to interact with non-AV road users as well. Another example that several researchers have discussed is the importance of how AVs interact with pedestrians near the crosswalk [48, 49].

It is important to be highlighted that most of the works done in the AD sectors are still in the proof of concepts or prototypes stage [6, 50]. For mass production of AV platforms, to assure that the AD software is always operating to the desired level safely, the calibration topic for the sensors used on the vehicle should be focused on too. Companies such as Aurora Labs and Applied Intuition are founded to focus on software-defined vehicle-related diagnostics topics [51, 52].

Furthermore, to assure the safety of AD, topics such as AV remote monitoring are also studied and developed. For example, companies like Ottopia, Einride, and Sensible 4 are among the organizations which develop the said technology to yield safer driverless navigation on the road [53–55].

3.3. Vision and Current State of the AV Industry

As we have understood from the high-level description of how the AD stack is working, the next question is how is the technology going to be productized and marketed? Is it by having millions of decentralized AVs on the road owned by different owners? Or is it as a shared driverless future mobility? The discussions are ongoing [56–58]. However, out of all the visions on how future mobility should be, the shared driverless mobility perspectives that are brought by ACES mobility are frequently discussed. With this vision, AD will spur a new type of mobility with other emerging automotive elements in ACES mobility [59–61].

This has sparked the call to action which has been answered by cities, industries as well as research institutes. KPMG in its report called "Autonomous Vehicles Readiness Index" has indicated that the race toward CV and AV technology is not only among automotive companies but also cities [62]. In "The City Adaptation to the Autonomous Vehicles Implementation: Reimagining the Dubai City of Tomorrow" [63], the authors explicitly mentioned the vision of AVs in Dubai, one of the leading cities in the AV industry aimed to have 25% of its transportation be driverless by 2030 [64]. Consequently, a lot of companies such as Cruise and Jaguar Land Rover have been testing their AD technology in the city (Figure 3.9).

Norway, one of the "Shangri-la" for future mobility issues is also actively doing pilots for AV technology. For example, in one of the public projects by the European Union, FABULOS (Future Automated Bus Urban Level Operation Systems), several industrial consortiums carry out their self-driving technology trials in Gjesdal [65]. The findings from these pilots help contribute to the improvements of the technology.

According to Gartner Inc.'s annual report titled "Hype Cycle," as of July 2020, AV has entered the "trough of disillusionment" stage. Consequently, compared to the "hyped" expectation for the

FIGURE 3.9 AVs as the future mobility.

petovarga/Shutterstock.com.

technology in the 2010s, the expectation toward the AV industry is now getting more "sensible" and "realistic." This is because of the realization that the successful deployment of AVs in the mass market as part of ACES requires a lot of dependencies and cross-industrial collaborations [66, 67, 68].

Nevertheless, this does not mean that the industrial growth of AVs is canceled. Instead, the recent announcement of the partnership of major companies from different sectors to work on AVs showed the vast potential of this field [69, 70]. However, as can be seen from the overview of this chapter, much work, which involve collaborations in different areas such as technical, governmental, societal, and policymaking, needs to be done before AVs can be safely deployed to the general public.

3.4. Summary

One of the major motivations for vehicle automation R&D is its potential to facilitate the reduction of accidents. However, for AVs to reliably deliver a safe and dependable driverless maneuver, it is not as easy as navigating from point A to B. In this chapter, the author provides a technical overview of AD technology intended for general readers. As can be seen, manifesting AVs for widespread deployment is not a simple task. A lot of collaborations from different perspectives are required before the vision turns into reality. Therefore, it is hoped that this chapter provides a comprehensive overview of the complexities of the AV field to prevent a myopic view on the subject and therefore prevent irresponsible marketing of the technology before it has reached the required readiness. In the next chapter, we will provide an overview of the next element of the ACES technology, i.e., CVs.

References

1. Laprise, S. and Winrich, C., "The Impact of Science Fiction Films on Student Interest in Science," *Journal of College Science Teaching* 40, no. 2 (2010): 45.

2. Braun, R., "Autonomous Vehicles: From Science Fiction to Sustainable Future," in Aguiar, M., Mathieson, C., and Pearce, L. (eds.), *Mobilities, Literature, Culture* (Cham: Palgrave Macmillan, 2019), 259-280.

3. Jameson, F., *Archaeologies of the Future: The Desire Called Utopia and Other Science Fictions* (London: Verso, 2005)

4. Gandia, R.M., Antonialli, F., Cavazza, B.H., Neto, A.M. et al., "Autonomous Vehicles: Scientometric and Bibliometric Review," *Transport Reviews* 39, no. 1 (2019): 9-28.

5. Llorca, D.F., "From Driving Automation Systems to Autonomous Vehicles: Clarifying the Terminology," arXiv preprint arXiv:2103.10844, 2021.

6. Hamid, U.Z.A., Mehndiratta, M., and Adali, E., "Adopting Aviation Safety Knowledge into the Discussions of Safe Implementation of Connected and Autonomous Road Vehicles," SAE Technical Paper 2021-01-0074, 2021, https://doi.org/10.4271/2021-01-0074

7. Park, H., Khattak, Z., and Smith, B., "Glossary of Connected and Automated Vehicle Terms," 2018, accessed August 2022, http://www.ctb.virginia.gov/resources/2018/oct/tech/glossary-of-cav-terms-ver1.0-03052018-1.pdf

8. SAE, "SAE International Releases Updated Visual Chart for Its 'Levels of Driving Automation' Standard for Self-Driving Vehicles," accessed August 2022, https://www.sae.org/news/press-room/2018/12/sae-international-releases-updated-visual-chart-for-its-%E2%80%9Clevels-of-driving-automation%E2%80%9D-standard-for-self-driving-vehicles

9. SAE International, "Taxonomy and Definitions for Terms Related to Driving Automation Systems for On-Road Motor Vehicles," accessed August 2022, https://www.sae.org/standards/content/j3016_202104

10. Chui, M., Manyika, J., and Miremadi, M., "Four Fundamentals of Workplace Automation," *McKinsey Quarterly* 29, no. 3 (2015): 1-9.

11. Hamid, U.Z.A. and Al-Turjman, F. (eds.), "Introductory Chapter: A Brief Overview of Autonomous, Connected, Electric and Shared (ACES) Vehicles as the Future of Mobility," *Towards Connected and Autonomous Vehicle Highways* (Cham: Springer, 2021), 3-8.

12. Cummings, M., Li, S., Seth, D., and Seong, M., "Concepts of Operations for Autonomous Vehicle Dispatch Operations," No. CSCRS-R9, Collaborative Sciences Center for Road Safety, 2021.

13. SAE International, "SAE International Journal of Connected and Automated Vehicles 2021 Special Issue: Robust, Safe, and Secure Implementation of Connected and Automated Vehicles' Functional Software Design and Architecture: Facilitating a Reliable, Feasible, and Comfortable Future Mobility," accessed August 2022, https://www.sae.org/publications/collections/content/jrn-ca-si-04/

14. Hamid, U.Z.A., Sezer, V., Li, B., Huang, Y. et al. (eds.), "Introductory Chapter: Roles of Path Planning in Providing Reliable Navigation and Control for Autonomous Vehicles and Robots," in *Path Planning for Autonomous Vehicles-Ensuring Reliable Driverless Navigation and Control Maneuver* (London: IntechOpen, 2019).

15. Hamid, U.Z.A., Kyyhkynen, A., Peralta-Cabezas, J.L., Saarinen, J. et al., "All-Weather Autonomous Vehicle: Performance Analysis of an Automated Heavy Quadricycle in Non-Snow and Snowstorm Conditions Using Single Map," in *The IAVSD International Symposium on Dynamics of Vehicles on Roads and Tracks* (Cham: Springer, 2019), 1100-1106.

16. Kuutti, S., Fallah, S., Katsaros, K., Dianati, M. et al., "A Survey of the State-of-the-Art Localization Techniques and Their Potentials for Autonomous Vehicle Applications," *IEEE Internet of Things Journal* 5, no. 2 (2018): 829-846.

17. Güzel, M.S., "Autonomous Vehicle Navigation Using Vision and Mapless Strategies: A Survey," *Advances in Mechanical Engineering* 5 (2013): 234747.

18. Agrawal, M., Konolige, K., and Bolles, R.C., "Localization and Mapping for Autonomous Navigation in Outdoor Terrains: A Stereo Vision Approach," in *2007 IEEE Workshop on Applications of Computer Vision (WACV'07)*, Austin, TX, 7, IEEE, February 2007.

19. Brummelen, V., Jessica, M.O.B., Gruyer, D., and Najjaran, H., "Autonomous Vehicle Perception: The Technology of Today and Tomorrow," *Transportation Research Part C: Emerging Technologies* 89 (2018): 384-406.

20. Al Nuaimi, K. and Kamel, H., "A Survey of Indoor Positioning Systems and Algorithms," in *2011 International Conference on Innovations in Information Technology*, Abu Dhabi, UAE, 185-190, IEEE, 2011.

21. Taketomi, T., Uchiyama, H., and Ikeda, S., "Visual SLAM Algorithms: A Survey from 2010 to 2016," *IPSJ Transactions on Computer Vision and Applications* 9, no. 1 (2017): 1-11.

22. Huang, B., Zhao, J., and Liu, J., "A Survey of Simultaneous Localization and Mapping," arXiv preprint arXiv:1909.05214, 2019.

23. Zhang, Y., Carballo, A., Yang, H., and Takeda, K., "Autonomous Driving in Adverse Weather Conditions: A Survey," arXiv preprint arXiv:2112.08936, 2021.

24. Llamazares, Á., Molinos, E.J., and Ocaña, M., "Detection and Tracking of Moving Obstacles (DATMO): A Review," *Robotica* 38, no. 5 (2020): 761-774.

25. Hadi, R.A., Sulong, G., and George, L.E., "Vehicle Detection and Tracking Techniques: A Concise Review," arXiv preprint arXiv:1410.5894, 2014.

26. Verma, R., "A Review of Object Detection and Tracking Methods," *International Journal of Advance Engineering and Research Development* 4, no. 10 (2017): 569-578.

27. Aryal, M., "Object Detection, Classification, and Tracking for Autonomous Vehicle," Doctoral dissertation, Grand Valley States University, 2018.

28. Tabone, W., De Winter, J., Ackermann, C., Bärgman, J. et al., "Vulnerable Road Users and the Coming Wave of Automated Vehicles: Expert Perspectives," *Transportation Research Interdisciplinary Perspectives* 9 (2021): 100293.

29. Yu, H., Yu, Z., Ting, L., and Sheng, L., "Topic Detection and Tracking Review," *Journal of Chinese Information Processing* 21, no. 6 (2007): 71-87.

30. Sun, Z., Bebis, G., and Miller, R., "On-Road Vehicle Detection: A Review," *IEEE Transactions on Pattern Analysis and Machine Intelligence* , no. 5 (2006): 694-628, 711.

31. Ravindran, R., Santora, M.J., and Jamali, M.M., "Multi-object Detection and Tracking, Based on DNN, for Autonomous Vehicles: A Review," *IEEE Sensors Journal* 21, no. 5 (2020): 5668-5677.

32. Mukhtar, A., Xia, L., and Tang, T.B., "Vehicle Detection Techniques for Collision Avoidance Systems: A Review," *IEEE Transactions on Intelligent Transportation Systems* 16, no. 5 (2015): 2318-2338.

33. Bezai, N.E., Medjdoub, B., Al-Habaibeh, A., Chalal, M.L. et al., "Future Cities and Autonomous Vehicles: Analysis of the Barriers to Full Adoption," *Energy and Built Environment* 2, no. 1 (2021): 65-81.

34. Hamid, U.Z.A., Saito, Y., Zamzuri, H., Rahman, M.A.A. et al., "A Review on Threat Assessment, Path Planning and Path Tracking Strategies for Collision Avoidance Systems of Autonomous Vehicles," *International Journal of Vehicle Autonomous Systems* 14, no. 2 (2018): 134-169.

35. Hamid, U.Z.A., Zamzuri, H., Rahman, M.A.A., Saito, Y. et al., "Collision Avoidance System Using Artificial Potential Field and Nonlinear Model Predictive Control: A Case Study of Intersection Collisions with Sudden Appearing Moving Vehicles," in Spiryagin, M., Gordon, T., Cole, C., and McSweeney, T. (eds.), *The Dynamics of Vehicles on Roads and Tracks* (Boca Raton, FL: CRC Press, 2017), 367-372.

36. Hamid, U.Z.A., "Vehicle Collision Avoidance for the Presence of Uncertain Obstacles Using Integrated Nonlinear Controller," 2018.

37. Li, Y., Li, K., Yang, Z., Morys, B. et al., "Threat Assessment Techniques in Intelligent Vehicles: A Comparative Survey," *IEEE Intelligent Transportation Systems Magazine* 13, no. 4 (2020): 71-91.

38. Magdici, S. and Althoff, M., "Adaptive Cruise Control with Safety Guarantees for Autonomous Vehicles," *IFAC-PapersOnLine* 50, no. 1 (2017): 5774-5781.

39. Sugihara, K. and Smith, J., "Genetic Algorithms for Adaptive Motion Planning of an Autonomous Mobile Robot," in *Proceedings 1997 IEEE International Symposium on Computational Intelligence in Robotics and Automation CIRA'97. "Towards New Computational Principles for Robotics and Automation"*, Monterey, Canada, 138-143, IEEE, 1997.

40. González, J.V.G., "Fast Marching Methods in Path and Motion Planning: Improvements and High-Level Applications," PhD dissertation, Universidad Carlos III de Madrid, 2015.

41. Katrakazas, C., Quddus, M., Chen, W.-H., and Deka, L., "Real-Time Motion Planning Methods for Autonomous On-Road Driving: State-of-the-Art and Future Research Directions," *Transportation Research Part C: Emerging Technologies* 60 (2015): 416-442.

42. Lefèvre, S., Vasquez, D., and Laugier, C., "A Survey on Motion Prediction and Risk Assessment for Intelligent Vehicles," *ROBOMECH Journal* 1, no. 1 (2014): 1-14.

43. Claussmann, L., Revilloud, M., Gruyer, D., and Glaser, S., "A Review of Motion Planning for Highway Autonomous Driving," *IEEE Transactions on Intelligent Transportation Systems* 21, no. 5 (2019): 1826-1848.

44. Viswanath, P., Mody, M., Nagori, S., Jones, J. et al., "Virtual Simulation Platforms for Automated Driving: Key Care-About and Usage Model," *Electronic Imaging* 2018, no. 17 (2018): 164-161.

45. Wang, Z., Bian, Y., Shladover, S.E., Wu, G. et al., "A Survey on Cooperative Longitudinal Motion Control of Multiple Connected and Automated Vehicles," *IEEE Intelligent Transportation Systems Magazine* 12, no. 1 (2019): 4-24.

46. Klomp, M., Jonasson, M., Laine, L., Henderson, L. et al., "Trends in Vehicle Motion Control for Automated Driving on Public Roads," *Vehicle System Dynamics* 57, no. 7 (2019): 1028-1061.

47. Iskander, J., Attia, M., Saleh, K., Nahavandi, D. et al., "From Car Sickness to Autonomous Car Sickness: A Review," *Transportation Research Part F: Traffic Psychology and Behaviour* 62 (2019): 716-726.

48. Millard-Ball, A., "Pedestrians, Autonomous Vehicles, and Cities," *Journal of Planning Education and Research* 38, no. 1 (2018): 6-12.

49. Rasouli, A. and Tsotsos, J.K., "Autonomous Vehicles that Interact with Pedestrians: A Survey of Theory and Practice," *IEEE Transactions on Intelligent Transportation Systems* 21, no. 3 (2019): 900-918.

50. Hamid, U.Z.A., Irimescu, D.S., and Zaman, M.T., "Challenges of Complex Software Development for Emerging Technologies in Automotive Industry: Bridging the Gap of Knowledge between the Industry Practitioners," SAE Technical Paper 2022-01-0109, 2022, https://doi.org/10.4271/2022-01-0109

51. Aurora Labs' Website, "Vehicle Software Intelligence: Solving the Challenge of Automotive Software Development," accessed August 2022, https://www.auroralabs.com

52. Applied Intuition's Website, "Software Solutions for Automotive," accessed August 2022, https://www.appliedintuition.com

53. Ottopia's Website, "Enabling Autonomy," accessed August 2022, https://ottopia.tech

54. Einride's Website, "Intelligent Movement," accessed August 2022, https://www.einride.tech/

55. Kotilainen, I., Händel, C., Hamid, U.Z.A., Santamala, H. et al., "Arctic Challenge Project's Final Report: Road Tranport Automation in Snowy and Icy Conditions," Väyläviraston tutkimuksia, 2019.

56. Pearl, T.H., "Compensation at the Crossroads: Autonomous Vehicles and Alternative Victim Compensation Schemes," in *AIES '19: AAAI/ACM Conference on AI, Ethics, and Society*, Honolulu HI, 187-193, 2019.

57. Lavasani, M., Jin, X., and Yiman, D., "Market Penetration Model for Autonomous Vehicles on the Basis of Earlier Technology Adoption Experience," *Transportation Research Record* 2597, no. 1 (2016): 67-74.

58. Attias, D. (ed.), "The Autonomous Car, a Disruptive Business Model?," in *The Automobile Revolution* (Cham: Springer, 2017), 99-113.

59. Modi, S., Spubler, A., and Jin, J., *Impact of Automated, Connected, Electric, and Shared (ACES) Vehicles on Design, Materials, Manufacturing and Business Models* (Ann Arbor, MI: Center for Automotive Research, 2018)

60. Bridges, F., "ACES: Autonomous, Connected, Electric and Shared Will Continue to Matter in Mobility in 2020, Says McKinsey & Company," 2020.

61. Möller, T., Padhi, A., Pinner, D., and Tschiesner, A., "The Future of Mobility Is at Our Doorstep," McKinsey Center for Future Mobility, 2019.

62. KPMG, "Autonomous Vehicles Readiness Index: Assessing Countries' Preparedness for Autonomous Vehicles," 2019.

63. Hafiz, D. and Zohdy, I., "The City Adaptation to the Autonomous Vehicles Implementation: Reimagining the Dubai City of Tomorrow," in Hamid, U.Z.A. and Al-Turjman, F. (eds.), *Towards Connected and Autonomous Vehicle Highways* (Cham: Springer, 2021), 27-41.

64. Shift, "Dubai Wants 25% of All Journeys in the City to Be 'Driverless' by 2030," TheNextWeb, accessed August 2022, https://thenextweb.com/news/dubai-25-percent-driverless-2030-syndication

65. FABULOS Project, "GJESDAL PILOT 2," accessed August 2022, https://fabulos.eu/gjesdal-pilot-2/

66. Visnic, B., "2020 Hype Cycle for Connected and Smart Mobility," SAE International, accessed August 2022, https://www.sae.org/news/2020/09/2020-hype-cycle-for-connected-vehicles-and-smart-mobility

67. Elsden, M., "Gartner Hype Cycle 2020 for Connected Autonomous Vehicles (CAV) and Smart Mobility," LinkedIn Pulse, accessed August 2022, https://www.linkedin.com/pulse/gartner-hype-cycle-2020-connected-autonomous-vehicles-elsden-phd-mba

68. Wang, Y. and Sarkis, J., "Emerging Digitalisation Technologies in Freight Transport and Logistics: Current Trends and Future Directions," *Transportation Research Part E: Logistics and Transportation Review* 148 (2021): 102291.

69. Microsoft News Center, "Cruise and GM Team Up with Microsoft to Commercialize Self-Driving Vehicles," Microsoft, accessed August 2022, https://news.microsoft.com/2021/01/19/cruise-and-gm-team-up-with-microsoft-to-commercialize-self-driving-vehicles/

70. Jennings, R., " Apple Assembler Foxconn's Electric Vehicle Plans Start to Take Shape," *Forbes*, accessed August 2022, https://www.forbes.com/sites/ralphjennings/2021/10/21/apple-assembler-foxconns-electric-vehicle-plans-start-to-take-shape/

Connected Vehicles: Concise Overview

4.1. Background

Humans communicate because it is our nature. The action is carried out to share our thoughts, ideas, notions, and concepts with other parties [1]. Many benefits are achieved by communicating, such as security benefits from neighborhood watch activities stemmed from "communication" among people [2]. Communication and networking among humans can also bring benefits from an economic perspective, as demonstrated by social media platforms such as LinkedIn [3]. Similar things can be seen in the animal kingdom. The swarm concept in the ant colony, for example, brings a lot of benefits to the said organisms [4]. Therefore, it is evident that being connected brings a lot of benefits.

In recent decades, connectivity has been disrupted and is continuously transforming. One of the connectivity transformation timeline examples is the telephone. During the 1980s, the method of telephony was mostly the landline phone. Around the 1990s, mobile phones were widely adopted, followed by text messages via SMS (Short Message Service) gaining momentum thereafter. Fast forward, the widespread adoption of smartphones by the vast majority has sparked a new way of communication among humans [5] (Figure 4.1).

FIGURE 4.1 The telecommunication landscape has seen a lot of transformation in the last few decades.

Sasha Ka/Shutterstock.com.

The parallel and simultaneous rapid growth of the Internet in the meantime has facilitated and enabled data sharing between different platforms. The capacities of the data sharing size have also exponentially grown since then. The Internet catalyzed the Internet of Things (IoT), a technology phenomenon that allows everything to be connected to anything [6]. With the IoT, one can control his home lights with smart house management and monitor his devices remotely via the Internet [7]. Compared to other means of communication such as Bluetooth and RFID (Radio Frequency Identification), IoT provides better connectivity with more extensive data exchange. This has then sparked further discussions about the connected and Smart City [8].

According to Hamid et al. [8], a Smart City envisions a community where the inhabitants and the city are highly integrated and connected with both artificial and human intelligence. Smart City is one of the cornerstones of the FIR discussions, with IoT as one of the enablers. Consequently, the transformation will also disrupt the way humans communicate.

Smart City exemplified several main elements in the UN SDGs such as Affordable and Clean Energy; Industry, Innovation, and Infrastructure; and Sustainable Cities and Communities. This shows that market interest in Smart City–related businesses is expected to be worth more than 1 trillion USD by 2025, which includes the topic of CVs [9, 10].

Similar to the benefits of human communication and connectivity, many research and products show that communication among vehicles will help improve road safety as well as the quality of mobility [11, 12]. Accordingly, for ACES vehicles as the future mobility, connectivity between different shared driverless EVs will help facilitate a good AD experience. This is because a non-connected and decentralized AV mobility system will not yield the best benefits from the emerging technology such as solving traffic jams and reducing collisions, among many others. For ACES to be able to yield its full potential, AVs need to be connected, which will enable data and information sharing among different AVs. But the question is "What exactly is a connected vehicle?"

4.2. Connecting and Connected

Etymologically, Connected Vehicles (CVs) refer to vehicles that can interact and communicate with their environment or other external systems via communication channels. Vehicle connectivity allows vehicles to share data with other vehicles and other platforms, as well as receive data from those systems.

For clarity, when we discuss vehicle connectivity, there are two main perspectives that we need to consider. The first one is the elements or methodology to link and connect the vehicles and other external systems. For example, 5G technology is a broadband cellular network that has been studied for Connected and Automated Vehicle (CAV) technology integration. The second one, on the other hand, is the capabilities that can be devised when the vehicles are connected. Good vehicular connectivity will eventually enable true ACES implementations [6].

Since vehicular connectivity is a more mature and extensive topic than AVs, it is very tricky to share about CVs in a single go. Therefore, in this chapter, the author reviews several applications and topics related to the field of CVs and their relationship to ACES mobility. Through this review, readers are expected to gain a comprehensive understanding of the applications in the field of CVs. It is important to note that the discussions will be presented as an overview, and interested readers are advised to continue reading the cited references. Some of the following discussions might be similar to each other, but the author believes that by explaining these in different subsections, the readers will be able to grasp the subject better.

4.3. Back-End Overview

4.3.1. Vehicular Ad Hoc Network

Among the main factors and motivations for linking vehicles are safety and increased comfort. This has stimulated discussions on Vehicle-to-Vehicle (V2V) communication in the past few decades [13]. One of the main facilitators for the communication among vehicles, whether stationary or moving, is VANET, which stands for Vehicular Ad hoc Network. In this section, we will briefly try to understand the process of VANET.

VANET as it stands for is a network of communications for vehicles. This can be road vehicles such as cars and trucks. The network permits different types of vehicular communication, such as V2V communication [14, 15], vehicle-to-roadside communication [16], and inter-roadside communication. VANET is a structure that can utilize various types of communication technology. With the VANET framework, a lot of traffic and vehicular applications can be improved, for example, traffic road management, as well as road safety [17]. To enable constant connectivity, VANET is supported by Road Side Units, which are typically installed along roads, with the main purpose to assist the V2V and vehicle-to-roadside communication [18]. Therefore, it is implied that VANET has a dependence on infrastructure readiness as well.

VANET has sparked a lot of dialogues on the possibilities in the automotive and mobility sectors. For example, not only does it have the possibility to reduce road fatalities but it can also aid the police in detecting traffic violations [19]. Apart from that, a lot of companies have been investigating how to expand and improve VANET technology. dSPACE, for example, has been working to integrate VANET-based simulation with Hardware-in-the-Loop Testing. This will eventually allow for the future automotive technology to be validated and verified in the said simulated environment within real time, shortening the development time as well as time to market [20].

With the arrival of robotaxis, VANET potential has also risen [21]. This has been scintillated by other progress within the connectivity and telecommunications area such as 5G and Dedicated Short-Range Communications (DSRC). For ACES as the future mobility, improved communication channels need to be developed to support the vehicular network as the data that will be transferred by driverless vehicles are of larger volumes. Furthermore, the discussions on data privacy should also be emphasized to prevent the interference of transmissions between different vehicles during the driverless journey.

4.3.2. Cloud Technology

Future vehicles are no longer defined and constrained by their hardware specifications. Instead, with software-defined vehicles, the value of vehicles in the future will be defined by their software quality and the values that come with it [22]. This means automotive companies not only need to develop the technology behind the ACES vehicles, but they also need to make sure that the lifecycle of the software products of their vehicles is maintained properly. This includes the ability to provide over-the-air updates for ACES vehicle software without requiring the customers to physically visit the carmakers' sites. One of the enablers of this requirement is cloud technology, which interacts as a "middle-layer" of the communication between the software back-end management with the vehicle. Especially for AD platforms, this technology is particularly important in sending diagnostics data from the platform to the back-end engineers in the event of road accidents. More details on vehicular cloud technology can be found in the works by Odat and Ganesan [23], Banijamali et al. [24], and Khurram et al. [25]. Tech giants such as Microsoft and Amazon are also working on these topics, which suggest the importance of cloud technology for progress in ACES [26].

4.3.3. Dedicated Short-Range Communications

One of the most widely investigated vehicular wireless communications is DSRC. It enables V2V and vehicle-to-roadside communication without the need for cellular infrastructure [27]. DSRC is based on IEEE 802.11p, which provides the vehicle with a sharing platform to transfer securely its vehicle states, which include position, direction, and speed, to surrounding vehicles or infrastructure [28]. With this, multiple-vehicle collisions such as pileup incidents can be prevented by sharing the information about the frontal hazards in advance to the following vehicles, enhancing the ADAS and active safety performance of road vehicles [29]. The Basic Safety Message usually consists of vehicular motion and kinematics data such as its coordinates, speed, current actuators reading, and trajectory history [28, 30].

In the work by Bai et al. [31], the researchers from General Motors (GM) Global R&D and the North Carolina State University have observed and analyzed the performance of DSRC characteristics from different perspectives, such as data rate, vehicle mobility, reliability, and spatial, temporal, and symmetric correlations. The improvements of DSRC have been continuously explored, and it is expected to bring benefits to ACES implementations [32].

4.3.4. Cellular Vehicle-to-Everything (C-V2X)

If DSRC is based on IEEE 802.11p, C-V2X is a 3GPP (Third-Generation Partnership Project) standard for technology enabling vehicular communication. As the name suggests, the communication system is based on a mobile cellular connectivity method such as 4G or 5G for data exchange among vehicles and other network systems. C-V2X benefits from the utilization of input from direct V2V communication as well as network communications [33].

Companies like Qualcomm are working on the said technology, where they envisioned the transfer of data from infrastructure to the vehicle to improve the journey experience [34]. However, as it is based on wireless communication, among the drawbacks of C-V2X issues are security and privacy of the systems, which can expose the vehicle to the possibility of hacking. However, as discussed earlier in Chapter 3, there are indications that technologies such as C-V2X and DSRC will facilitate better collision warning systems, especially in hazardous scenarios involving occluded moving objects. Because of the high speed of C-V2X connectivity, it is expected to benefit safety-critical missions of ACES applications. However, as noted by Mannoni et al. [35], C-V2X performance depends on the user density and level of congestion. Therefore, more work should be done to improve the technical performance of the technology.

4.3.5. Low-Power Wide-Area Network

The standard Wide-Area Network (WAN) facilitates 3G, 4G, and 5G communications over the long distance. However, despite the fast rate, WAN consumes a lot of power [36]. For the ACES mobility use case, this performance is not adequate considering the vehicle will also be dependent on electricity. Therefore, high-power consumption could also increase range anxiety, preventing high user acceptance of the ACES vehicle technology [37].

This notion has motivated the study on Low-Power Wide-Area Network (LPWAN). One notable example is the Long-Range (LORA) module. LORA is a proprietary technology owned by SEMTECH, a technology company now based in Southern California [38]. According to its official website, LORA is capable of offering long-range Internet with low power. It can be used in a lot of IoT-based technology, which includes CVs. This technology will bring a lot of benefits in the context of ACES, especially for the sharing of data from sensors between different vehicles because of its long-range, low-power consumption.

4.3.6. 5G and Its Importance for Vehicle Connectivity

If one is following the telecommunication and networking industries, he or she might notice that in recent years, one of the most hyped buzzwords is 5G. What actually is 5G?

5G is a fifth-generation standard for cellular broadband network technology. The potential applications range from medical robotics [39] and mining industry [40] to safety-critical applications such as aviation and automotive [41, 42] (Figure 4.2).

FIGURE 4.2 5G is disrupting a lot of sectors, including automotive and mobility.

Suwin/Shutterstock.com.

Among the concerns with 4G for driverless mobility use cases, as can be shown in several R&D projects, are the network latency. With 5G, better connectivity and bandwidth are expected [43–45].

However, it should be noted that 5G itself is just one of the myriads of complex elements that will yield safe ACES mobility. Nevertheless, with better connectivity, connected driverless vehicles can be expected to share sensor information and data with much higher performance confidence. This can then be used for traffic management and collision avoidance of ACES vehicles.

4.3.7. A Glimpse of 6G, the Probable Next Step in the Vehicle Connectivity Field

Since we are already discussing 5G connectivity, it is considerable and thoughtful to mention the rapidly studied 6G. As the name suggests, 6G stands for the sixth-generation standard of cellular networks. Even though 5G itself is not yet widely employed and deployed, 6G is already examined. In Finland, a program called 6G Flagship is organized in the city of Oulu to study the potential impact of both 5G and 6G on future technologies [46].

For the automotive and mobility sectors, 6G has been positively examined in the past few years, with flagship vehicular technology journals such as *IEEE Vehicular Technology Magazine* organizing a couple of Special Issues on 6G topics [47]. According to He et al. [48], for 6G to enable the full potential of ACES technology, it should provide the driverless platform with network improvements on several aspects, which include (i) high reliability, (ii) low latency, (iii) massive access and scalability, and (iv) high throughput to deal with the massive amount of data during the AD navigation. It is still early to perceive the directions of 6G applications for ACES; however, it is something worth to be monitored [49].

4.4. Applications

4.4.1. Vehicle-to-Everything

We have talked about the connectivity approaches, systems, and processes in the previous sections; therefore, in the next subsections, we will look at several types of the most common implementations and applications of CVs. VANET is a network that enables communications between different vehicles; however, the communication between a vehicle with other entities, regardless of whether it is a vehicle or not, is called "Vehicle-to-Everything," also commonly known as V2X (Figure 4.3).

Among the popular types of V2X are Vehicle-to-Vehicle (V2V), Vehicle-to-Infrastructure (V2I), Vehicle-to-Grid, Vehicle-to-Cloud, and Vehicle-to-Pedestrian (V2P), among many others. The term has also been expanded into further perspectives such as business viewpoint. For example, Vehicle-to-Customer refers to vehicle communication with the customers, which eventually enables Vehicle-to-Building and Vehicle-to-Home concepts [50, 51].

With good connectivity, the vehicle will be able to communicate with anything via the IoV. The benefits are also enormous. For example, with V2V using DSRC, the ACES vehicle platform can

FIGURE 4.3   V2X enables vehicular communication with its surroundings and beyond.

Blue Planet Studio/Shutterstock.com.

share information about ongoing roadworks to the following vehicle. Therefore, the following vehicle will be able to replan its trajectory using the said information, thus preventing road accidents.

The safety of the pedestrian can also be assured with the V2P, especially during encounters with driverless vehicles at nighttime when visibility is limited. With the aid of V2P of the V2X technology, the ACES vehicle will be aware of the information of the surrounding environment, including those located in its blind spot. In summary, V2X will help improve the benefits of ACES for the mass audience. Interested readers are invited to read "Review of Research on V2X Technologies, Strategies, and Operations" [52], a comprehensive survey on V2X for further understanding.

4.4.2. Vehicle Connectivity for Improved Active Safety

As mentioned in Chapter 3, on a certain occasion, for example, in blind-spot monitoring, as well as potential collision scenario involving previously occluded objects, the information from local sensors on the vehicle alone are not sufficient. Therefore, integrating ACES technology with good vehicle connectivity technology will provide redundancy in the perception system. Hamid et al. [8] have shown the benefits of integrating vehicle connectivity into the AV software operation. The authors provided a simulation where an AV is moving on a road where suddenly undetected potholes appear on the trajectory of the AV. In Scenario 1, in the situation where the vehicle does not have the connectivity feature, the pothole is not avoided, therefore potentially damaging the vehicle. In Scenario 2 where the vehicle is equipped with connectivity, the information about the pothole can be shared with the host vehicle via the Internet. This information is then utilized by the AD stack to replan the trajectory, therefore enabling the obstacle avoidance maneuver.

Another use case to highlight the benefit of vehicle connectivity in terms of active safety improvements is "Road Weather Motorist Advisories and Warnings." According to the Intelligent Transportation Systems Joint Program Office of the US Department of Transportation [53], "Motorist Advisories and Warnings" works by providing warnings and information about bad weather to the vehicle. This will help the drivers increase their attention as well as driving quality. For driverless vehicles, this information can also be used to help the behavior planning modules to reduce the speed of the vehicle during bad weather, for example. It is evident from this subsection that vehicle connectivity has the vast potential of improving active safety for both conventional private cars and AVs.

4.4.3. Vehicle Platooning

In use cases, ACES mobility is not only limited to people mover functionality, but several companies such as Gatik in the USA and Einride in Sweden have been focusing their business model on the movement of goods [54, 55]. In this context, platooning of AVs is among those that must be taken into account during the development. In each AV platoon, each of the vehicles in the platoon needs to maintain a safe distance [56]. In addition to the local sensors from the environmental awareness modules of AVs, external information from other vehicles in the platoon will help ensure the safety maneuver of the vehicle fleet. For example, Wang et al. in their works have integrated V2X communication into their platooning strategy [57].

4.4.4. Improved Infotainment

In *Performability in Internet of Things* [8], the idea of IoV for improving infotainment in the vehicle is highlighted. As ACES is a disruptive innovation, it is expected that it will also transform the transportation sector. Therefore, with driverless vehicles, better infotainment is needed during AD navigation. Thus, one of the use cases of vehicle connectivity is to improve infotainment in the vehicle. With better Internet connection, passengers can be assured of a comfortable AD experience, particularly for a long-distance highway trip.

4.4.5. Improved GPS and Traffic Jam Reduction

For ACES to perform reliably, it needs to be able to locate itself with a very accurate and precise positioning strategy on a defined map. Vehicle connectivity is important for this purpose as it will be able to improve GPS information. Furthermore, a company called What3words has potentially transformed the positioning field by introducing its proprietary geocode system [58]. With the present invention, it will help both human-driven and driverless vehicles to locate themselves within a map with pinpoint accuracy (Figure 4.4).

Gridlock is a classic example of a bad traffic jam. In most countries, usually, traffic police members will be deployed to help improve the traffic [59]. However, this sometimes can cause additional stress to the police in the long run. With V2V and V2I technology, it is shown that the technology can help in improving the traffic jam as part of the urban traffic management of CVs [60]. With the shared data between vehicles and infrastructure, a new route can be suggested to the vehicles, therefore reducing the potential of traffic jams.

FIGURE 4.4 Vehicular technology will help in expediting the Smart ACES Highway.

jamesteohart/Shutterstock.com.

4.4.6. Vehicle Connectivity Enabling Automated Delivery

When we are looking at the grand vision of ACES, we will see that there will also be the potential for automated delivery, as mentioned in one of the previous subsections [54, 55]. Vehicle connectivity helps to improve particularly the last-mile delivery services industry with the ACES vehicles. The vehicle location can easily be tracked with good connectivity within the vehicle systems [61].

4.4.7. Vehicle Connectivity Improving Shared Mobility

Vehicle connectivity is also important in enabling the shared driverless mobility experience. With good connectivity within the vehicle system, the fleet management system software will be able to notify the user and potential passenger about the current information of the vehicle. This might also include the information of the passengers on the bus and the estimated time to the waiting point [62].

4.5. Vehicle Connectivity Roles in Enabling True ACES Mobility

AV development is not sustainable if it is only a standalone driverless vehicle amid human-driven vehicles. In fact, if that is the reality, it might cause more harm than good. Consequently, for AVs to transform the automotive industry and disrupt human living, subsequently achieving the objectives of UN SDGs and bringing the benefits economically and socially, it needs to be an element and part of the ACES mobility. From this chapter, it is apparent that vehicle connectivity will be able to link the AV with other elements in ACES, for example, shared mobility. ACES can also improve the active safety experience of the AD journey.

The following Figure 4.5 is drawn to illustrate the roles and contributions of vehicle connectivity to the ACES ecosystem, particularly in terms of improving the safety and shared mobility experience. In Zone I, the AV is driving in an urban area where it reaches a bumpy road because of ongoing roadwork constructions. This

FIGURE 4.5 Example of vehicle connectivity role in the ACES ecosystem [8, 63, 64].

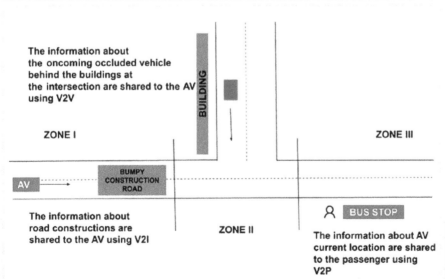

information cannot be detected by the local sensors on the AV. However, as that information is published from the infrastructure, the connected AV receives the information, therefore allowing it to reduce speed and prevent the uncomfortable ride on the bumpy road. In addition, in Zone II, the AV manages to avoid a collision with the oncoming occluded vehicle because its V2V feature sharing information about the other vehicle. When it reaches Zone III, its arrival information is sent to the mobile application of the passenger who is waiting at the bus stop. This is useful for the passenger as it helped them to plan their journey throughout the day. As can be seen, this is among the examples of how vehicle connectivity can play a major role in the ACES ecosystem. Further reading of the example in Figure 4.5 can be found in the works by Hamid et al. [8], Jurgen [63], and Heineke et al. [64].

4.6. Vision and Current State of the CV Industry

As the industry is now moving more toward the direction of CAM, the cybersecurity risks derived from the CVs application are getting more attention. The European Union Agency for Cybersecurity (ENISA), in their report "Recommendations for the Security of CAM" (in which the author is also part of the report), mentioned that one of the biggest challenges in addressing the cybersecurity issues for CAM is the lack of support from the top management in organizations [65]. Therefore, continuous works and regulations are now being promoted on the security and privacy of vehicle connectivity in different regions all around the world.

Furthermore, testbeds for CVs testing purposes have been developed all around the world. This marks the gravity of the topic, and governments and industries are starting to push this effort forward. Among the notable examples of CV testbeds are Korea Automobile Testing and Research Institute (Korea), Cooperative Vehicle Infrastructure System (CVIS, Tongji University, China), Australian Integrated Multimodal EcoSystem (AIMES, Australia), and Multi-Modal Intelligent Traffic Signal System (MMITSS) in

California and Arizona (USA). More details can be found in the work by Emami et al. [66].

Besides these efforts, companies and industry experts involved in the productization of vehicle connectivity in ACES mobility have extensively studied the topics of mobile edge computing for vehicles, networking for over-the-air software updates of CAV, and privacy and security [67–69].

It is clear that for ACES to be deployed safely to the mass audience, a lot of collaborations not only between different applications but also different industries are required.

4.7. Summary

In this chapter, the author provides a comprehensive overview and broad perspective of the CV topic. The survey is divided into two, i.e., the back end of the vehicle connectivity and the applications of CVs. Industrial trends and surveys are also concisely denoted. The author believes that this chapter will help the general reader understand the big picture of the role of CV in the ACES ecosystem.

References

1. Sperber, D., "How Do We Communicate," in Brockman, J. and Matson, K. (eds.), *How Things Are: A Science Toolkit for the Mind* (New York: Morrow, 1995), 191-199.

2. Pridmore, J., Mols, A., Wang, Y., and Holleman, F., "Keeping an Eye on the Neighbours: Police, Citizens, and Communication within Mobile Neighbourhood Crime Prevention Groups," *The Police Journal* 92, no. 2 (2019): 97-120.

3. Koch, T., Gerber, C., and De Klerk, J.J., "The Impact of Social Media on Recruitment: Are You LinkedIn?" *SA Journal of Human Resource Management* 16, no. 1 (2018): 1-14.

4. Kordon, A.K. (ed.), "Swarm Intelligence: The Benefits of Swarms," in *Applying Computational Intelligence* (Berlin, Heidelberg: Springer, 2010), 145-174.

5. Reed, J.H., Bernhard, J.T., and Park, J.-M., "Spectrum Access Technologies: The Past, the Present, and the Future," *Proceedings of the IEEE* 100, no. Special Centennial Issue (2012): 1676-1684.

6. Al-Turjman, F. (Eds), *Performability in Internet of Things* (Cham: Springer, 2018)

7. Balyk, N., Leshchuk, S., and Yatsenyak, D., "Developing a Mini Smart House Model," in *CEUR Workshop Proceedings*, Kryvyi Rih, Ukraine, November 29, 2019.

8. Hamid, U.Z.A., Zamzuri H., and Limbu D.K., "Internet of Vehicle (IoV) Applications in Expediting the Implementation of Smart Highway of Autonomous Vehicle: A Survey," in Al-Turjman, F. (ed.), *Performability in Internet of Things* (Cham: Springer, 2019), 137-157.

9. Anthopoulos, L.G. (ed.), "The Smart City Market," in *Understanding Smart Cities: A Tool for Smart Government or an Industrial Trick?* (Cham: Springer, 2017), 187-213.

10. Anthopoulos, L., Fitsilis, P., and Ziozias, C., "What Is the Source of Smart City Value?: A Business Model Analysis," in Information Resources Management Association (USA) (eds), *Smart Cities and Smart Spaces: Concepts, Methodologies, Tools, and Applications* (Hershey, PA: IGI Global, 2019), 56-77.

11. Michel, P., Karbowski, D., and Rousseau, A., "Impact of Connectivity and Automation on Vehicle Energy Use," SAE Technical Paper 2016-01-0152, 2016, https://doi.org/10.4271/2016-01-0152

12. Lempert, R.J., Preston, B., Charan, S.M., Fraade-Blanar, L. et al., "The Societal Benefits of Vehicle Connectivity," *Transportation Research Part D: Transport and Environment* 93 (2021): 102750.

13. Siegel, J.E., Erb, D.C., and Sarma, S.E., "A Survey of the Connected Vehicle Landscape—Architectures, Enabling Technologies, Applications, and Development Areas," *IEEE Transactions on Intelligent Transportation Systems* 19, no. 8 (2017): 2391-2406.

14. Ararat, O. and Aksun-Guvenc, B., "A Survey of Recent Developments in Collision Avoidance, Collision Warning and Inter-Vehicle Communication Systems," arXiv preprint arXiv:2012.12441, 2020.

15. Wang, J., Liu, J., and Kato, N., "Networking and Communications in Autonomous Driving: A Survey," *IEEE Communications Surveys & Tutorials* 21, no. 2 (2018): 1243-1274.

16. Wu, C., Yoshinaga, T., Ji, Y., and Zhang, Y., "Computational Intelligence Inspired Data Delivery for Vehicle-to-Roadside Communications," *IEEE Transactions on Vehicular Technology* 67, no. 12 (2018): 12038-12048.

17. Kiela, K., Barzdenas, V., Jurgo, M., Macaitis, V. et al., "Review of V2X-IoT Standards and Frameworks for ITS Applications," *Applied Sciences* 10, no. 12 (2020): 4314.

18. Xue, L., Yang, Y., and Dong, D., "Roadside Infrastructure Planning Scheme for the Urban Vehicular Networks," *Transportation Research Procedia* 25 (2017): 1380-1396.

19. Oche, M., Noor, R.M., Al-jawfi, A.S., Bimba, A.T. et al., "An Automatic Speed Violation Detection Framework for VANETs," in *2013 IEEE International Conference on RFID-Technologies and Applications (RFID-TA)*, Johar Bahru, Malaysia, 1-6, IEEE, 2013.

20. Buse, D.S., Schettler, M., Kothe, N., Reinold, P. et al., "Bridging worlds: Integrating Hardware-in-the-Loop Testing with Large-Scale VANET Simulation," in *2018 14th Annual Conference on Wireless On-Demand Network Systems and Services (WONS)*, Isola, France, 33-36. IEEE, 2018.

21. Vaidya, B. and Mouftah, H.T., "IoT Applications and Services for Connected and Autonomous Electric Vehicles," *Arabian Journal for Science and Engineering* 45, no. 4 (2020): 2559-2569.

22. Truong, N.B., Lee, G.M., and Ghamri-Doudane, Y., "Software Defined Networking-Based Vehicular Adhoc Network with Fog Computing," in *2015 IFIP/IEEE International Symposium on Integrated Network Management (IM)*, Ottawa, ON, Canada, 1202-1207, IEEE, 2015.

23. Odat, H.A. and Ganesan, S., "Firmware over the Air for Automotive, Fotamotive," in *IEEE International Conference on Electro/Information Technology*, Harbin, China, 130-139, IEEE, 2014.

24. Banijamali, A., Jamshidi, P., Kuvaja, P., and Oivo, M., "Kuksa: A Cloud-Native Architecture for Enabling Continuous Delivery in the Automotive Domain," in Franch, X., Männistö, T., Martínez-Fernández, S. (eds.), *International Conference on Product-Focused Software Process Improvement* (Cham: Springer, 2019), 455-472.

25. Khurram, M., Kumar, H., Chandak, A., Sarwade, V. et al., "Enhancing Connected Car Adoption: Security and Over the Air Update Framework," in *2016 IEEE 3rd World Forum on Internet of Things (WF-IoT)*, Reston, VA, 194-198, IEEE, 2016.

26. Deng, H.-W., Rahman, M., Chowdhury, M., Sabbir Salek, M. et al., "Commercial Cloud Computing for Connected Vehicle Applications in Transportation Cyber-Physical Systems," arXiv preprint arXiv:2008.07290, 2020.

27. Abboud, K., Omar, H.A., and Zhuang, W., "Interworking of DSRC and Cellular Network Technologies for V2X Communications: A Survey," *IEEE Transactions on Vehicular Technology* 65, no. 12 (2016): 9457-9470.

28. Bansal, G., Kenney, J.B., and Rohrs, C.E., "LIMERIC: A Linear Adaptive Message Rate Algorithm for DSRC Congestion Control," *IEEE Transactions on Vehicular Technology* 62, no. 9 (2013): 4182-4197.

29. Shrestha, R., Bajracharya, R., and Nam, S.Y., "Challenges of Future VANET and Cloud-Based Approaches," *Wireless Communications and Mobile Computing* 2018 (2018): 1-15.

30. Kenney, J.B., "Dedicated Short-Range Communications (DSRC) Standards in the United States," *Proceedings of the IEEE* 99, no. 7 (2011): 1162-1182.

31. Bai, F., Stancil, D.D., and Krishnan, H., "Toward Understanding Characteristics of Dedicated Short Range Communications (DSRC) from a Perspective of Vehicular Network Engineers," in *Proceedings of the Sixteenth Annual International Conference on Mobile Computing and Networking*, Chicago, IL, 329-340, 2010.

32. Wu, J., Zhang, L., and Liu, Y., "On the Design and Implementation of a Real-Time Testbed for Distributed TDMA-Based MAC Protocols in VANETs," *IEEE Access* 9 (2021): 122092-122106.

33. Gyawali, S., Xu, S., Qian, Y., and Hu, R.Q., "Challenges and Solutions for Cellular Based V2X Communications," *IEEE Communications Surveys & Tutorials* 23, no. 1 (2020): 222-255.

34. Bera, R., "Smart Automotive System with CV2X-Based Ad Hoc Communication," in Singh, G., Jain, V., Chatterjee, J.M., and Gaur, L. (eds.), *Cloud and IoT-Based Vehicular Ad Hoc Networks* (Hoboken, NJ: John Wiley & Sons, Inc., 2021), 293-323.

35. Mannoni, V., Berg, V., Sesia, S., and Perraud, E., "A Comparison of the V2X Communication Systems: ITS-G5 and C-V2X," in *2019 IEEE 89th Vehicular Technology Conference (VTC2019-Spring)*, Kuala Lumpur, Malaysia, 1-5, IEEE, 2019.

36. de Carvalho Silva, J., Rodrigues, J.J.P.C., Alberti, A.M., Solic, P. et al., "LoRaWAN—A Low Power WAN Protocol for Internet of Things: A Review and Opportunities," in *2017 2nd International Multidisciplinary Conference on Computer and Energy Science (SpliTech)*, Split, Croatia, 1-6, IEEE, 2017.

37. Neubauer, J. and Wood, E., "The Impact of Range Anxiety and Home, Workplace, and Public Charging Infrastructure on Simulated Battery Electric Vehicle Lifetime Utility," *Journal of Power Sources* 257 (2014): 12-20.

38. Semtech, "Semtech LoRa Technology Overview," accessed August 2022, https://www.semtech.com/lora

39. Soldani, D., Fadini, F., Rasanen, H., Duran, J. et al., "5G Mobile Systems for Healthcare," in *2017 IEEE 85th Vehicular Technology Conference (VTC Spring)*, Sydney, Australia, 1-5, IEEE, 2017.

40. Gaber, T., El Jazouli, Y., Eldesouky, E., and Ali, A., "Autonomous Haulage Systems in the Mining Industry: Cybersecurity, Communication and Safety Issues and Challenges," *Electronics* 10, no. 11 (2021): 1357.

41. Matti, E., Johns, O., Khan, S., Gurtov, A. et al., "Aviation Scenarios for 5G and Beyond," in *2020 AIAA/IEEE 39th Digital Avionics Systems Conference (DASC)*, San Antonio, TX, 1-10, IEEE, 2020.

42. Järvinen, J., "GACHA-robottibussin palvelumuotoilu: Nokia Karamalmi," 2020.

43. Kotilainen, I., Händel, C., Hamid, U.Z.A., Nykänen, L. et al., "Connected and Automated Driving in Snowy and Icy Conditions-Results of Four Field-Testing Activities Carried Out in Finland," *SAE Intl. J CAV* 4, no. 1 (2021): 109-118, https://doi.org/10.4271/12-04-01-0009

44. Kotilainen, I., Händel, C., Hamid, U.Z.A., Santamala, H. et al., "Arctic Challenge Project's Final Report: Road Transport Automation in Snowy and Icy Conditions," Väyläviraston tutkimuksia, 2019.

45. Tang, Z. and He, J., "NOMA Enhanced 5G Distributed Vehicle to Vehicle Communication for Connected Autonomous Vehicles," in *Proceedings of the ACM MobiArch 2020 the 15th Workshop on Mobility in the Evolving Internet Architecture*, London, UK, 42-47, 2020.

46. University of Oulu, "6G Flagship," accessed August 2022, https://www.oulu.fi/6gflagship/

47. IEEE VTS, "Call For Papers: 6G: What Is Next?," *IEEE Vehicular Technology Magazine*, accessed August 2022, http://www.ieeevtc.org/vtmagazine/specisu--6G.php

48. He, J., Yang, K., and Chen, H.-H., "6G Cellular Networks and Connected Autonomous Vehicles," *IEEE Network* 35, no. 4 (2020): 255-261.

49. Giordani, M., Polese, M., Mezzavilla, M., Rangan, S. et al., "Toward 6G Networks: Use Cases and Technologies," *IEEE Communications Magazine* 58, no. 3 (2020): 55-61.

50. Gschwendtner, C., Sinsel, S.R., and Stephan, A., "Vehicle-to-X (V2X) Implementation: An Overview of Predominate Trial Configurations and Technical, Social and Regulatory Challenges," *Renewable and Sustainable Energy Reviews* 145 (2021): 110977.

51. Siemens, "Vehicle-to-X (V2X) Communication Ttechnology," accessed August 2022, https://assets.new.siemens.com/siemens/assets/api/uuid:9c7f02efa4cd2e1b0f6ea0eade b5db658837d86e/siemens-vehicle-to-x-communication-technology-infographic.pdf

52. Pearre, N.S. and Ribberink, H., "Review of Research on V2X Technologies, Strategies, and Operations," *Renewable and Sustainable Energy Reviews* 105 (2019): 61-70.

53. United States Department of Transportation, "Road Weather—Motorist Advisories and Warnings," accessed August 2022, https://www.its.dot.gov/infographs/motorist_ advisories_warnings.htm

54. Mahdavian, A., Shojaei, A., Mccormick, S., Papandreou, T. et al., "Drivers and Barriers to Implementation of Connected, Automated, Shared, and Electric Vehicles: An Agenda for Future Research," *IEEE Access* 9 (2021): 22195-22213.

55. Orii, L., Tosca, D., Kun, A.L., and Shaer, O., "Perceptions on the Future of Automation in r/Truckers," in *Extended Abstracts of the 2021 CHI Conference on Human Factors in Computing Systems*, Yokohama Japan, 1-6, May 2021.

56. Ariffin, M.H.M., Rahman, M.A.A., and Zamzuri, H., "Effect of Leader Information Broadcasted throughout Vehicle Platoon in a Constant Spacing Policy," in *2015 IEEE International Symposium on Robotics and Intelligent Sensors (IRIS)*, Langkawi, Malaysia, 132-137, IEEE, 2015.

57. Wang, P., Di, B., Zhang, H., Bian, K. et al., "Platoon Cooperation in Cellular V2X Networks for 5G and Beyond," *IEEE Transactions on Wireless Communications* 18, no. 8 (2019): 3919-3932.

58. Jiang, W. and Stefanakis, E., "What3Words Geocoding Extensions," *Journal of Geovisualization and Spatial Analysis* 2, no. 1 (2018): 1-18.

59. Kumar, G.R. and Mohan, S.R., "Work Stress for Traffic Police in Chennai City," *Journal of Contemporary Research in Management* 4, no. 2 (2009): 107-115.

60. Bento, L.C., Parafita, R., and Nunes, U., "Intelligent Traffic Management at Intersections Supported by V2V and V2I Communications," in *2012 15th International IEEE Conference on Intelligent Transportation Systems*, Anchorage, AK, 1495-1502, IEEE, 2012.

61. Abdelkader, G., Elgazzar, K., and Khamis, A., "Connected Vehicles: Technology Review, State of the Art, Challenges and Opportunities," *Sensors* 21, no. 22 (2021): 7712.

62. Zeng, X., Balke, K., and Songchitruksa, P., *Potential Connected Vehicle Applications to Enhance Mobility, Safety, and Environmental Security* (College Station, TX: Texas Transportation Institute, 2012)

63. Jurgen, R., *V2V/V2I Communications for Improved Road Safety and Efficiency* (Warrendale, PA: SAE, 2012)

64. Heineke, K., Ménard, A., Södergren, F., and Wrulich, M., "Development in the Mobility Technology Ecosystem—How Can 5G Help," McKinsey and Company, 2019.

65. European Union Agency for Cybersecurity, "Recommendations for the Security of CAM," ENISA, accessed August 2022, https://www.enisa.europa.eu/publications/recommendations-for-the-security-of-cam

66. Emami, A., Sarvi, M., and Asadi Bagloee, S., "A Review of the Critical Elements and Development of Real-World Connected Vehicle Testbeds around the World," *Transportation Letters* 14, no. 1 (2020): 49-74, doi:10.1080/19427867.2020.1759852

67. Ye, F., Guo, J., Kim, K.J., Orlik, P.V. et al., "Bi-level Optimal Edge Computing Model for On-Ramp Merging in Connected Vehicle Environment," in *2019 IEEE Intelligent Vehicles Symposium (IV)*, Paris, France, 2005-2011, IEEE, 2019.

68. Halder, S., Ghosal, A., and Conti, M., "Secure Over-the-Air Software Updates in Connected Vehicles: A Survey," *Computer Networks* 178 (2020): 107343.

69. He, Q., Meng, X., and Qu, R., "Survey on Cyber Security of CAV," in *2017 Forum on Cooperative Positioning and Service (CPGPS)*, Harbin, China, 351-354, IEEE, 2017.

5

Electric Vehicles: Concise Overview

5.1. Background

The Internal Combustion Engine (ICE) has been one of the backbones of modern transportation for more than a century, where, according to Rațiu [1], it was designed as early as the 1680s (Figure 5.1). Since then, it has benefitted and been utilized by different means of transportation such as road vehicles, buses, trucks, and aircraft as well as maritime purposes. The ICE facilitates the vehicle trip with the use of and fueled by natural resources such as gasoline and diesel [2].

ICE remains the undisputed main power resource for most transportation platforms ever since its introduction. One of the factors contributing to this claim is the abundance of petroleum resources [3]. Moreover, this is supported by the abundance of gas stations around the globe. Therefore, from an economic perspective, fuel-based ICEs are an affordable solution for private users and transportation companies.

Despite the interest in EVs since the 1830s, but because of the absence of the factors mentioned above for EVs (which prompted the wide acceptance of ICE), EV technology has yet to gain momentum in the market [4, 5]. However, in recent decades, awareness of the emission volumes produced by ICE-based vehicles has increased worldwide. This has inspired more R&D efforts and works on alternatives to replace the ICE, which include EVs.

FIGURE 5.1 ICE as the main backbone of vehicles, unchallenged until recent years.

PGMart/Shutterstock.com.

Several aspects influence the echoing acceptance of vehicle electrification. These include the rise of Millennials and Generation Z—new demographics of the world population—with different mindsets of customer behavior [6] (Figure 5.2). Furthermore, the FIR has also fueled the growth of the EV industry as can be seen in the birth and entry of a lot of start-ups into the market for the said topic since the 2010s [7]. With the UN SDGs, this topic has been intercommunicated with more aggressive and stronger tones. Sustainability has also been one of the hallmarks for the marketing of emerging technologies in recent years [8].

It should also be noted that EVs are not the only domain trying to help reduce the emissions caused by conventional ICE in recent years. In fact some works have been done in ICE-related areas to help in emissions reduction, such as emulsified fuels, which see the emulsification of combustible natural resources with water [9].

FIGURE 5.2 Millennials and Generation Z are prompting a lot of emerging disruptive technologies.

GENERATION CONCEPTS

Baby boomers **Generation X** **Millenials** **Gen Z**

However, EVs are getting more attention and focus because of several factors. Vehicle electrification-themed business activities by electricity companies like Fortum and Helen Ventures in Europe have signaled the huge market potential for EVs [10, 11]. Apart from that, electrification progress also encouraged a lot of entrepreneurial activities on electrification topics across different industries with companies such as Cake producing electric motorbikes, Candela with electric boats, and a company called Arrival building electric road vehicles in the United Kingdom (UK) [12–14] (Figure 5.3).

FIGURE 5.3 Electrification-themed development happening simultaneously within the different modes of transportation in recent years.

It is not a coincidence that major carmakers like GM and Bayerische Motoren Werke AG (BMW) are organizing massive "celebrity endorsement" marketing campaigns by the world's biggest superstars such as Will Ferrell, Mike Myers, and Arnold Schwarzenegger for their respective EVs. This is not a coincidence that can be overlooked. Instead it shows that the companies are willing to allocate a large amount of money into their marketing budget, implying how big the EV market will soon be [15–17].

But despite all this, public awareness and understanding of this topic are not yet high, which in turn creates range anxiety among potential users, which could hinder the widespread adoption of EVs [18]. Therefore, to overcome this, continuous education and information dissemination should be carried out to clarify the benefits and backgrounds of the technology to the public. Thus, the real question is "What exactly is an Electric Vehicle?"

5.2. Electrifying the World—The Motivation

What is electrification? Why do we need it? Terminologically, electrification refers to the effort to transform fossil-fuel-based technology into an electricity-based platform [19]. Electrification of vehicles is said to reduce the total cost of ownership as well as environmental pollution [20]. Compared to ICE-based vehicle drivetrains, EVs also have greater potential to reduce the impact of climate change because of their non-dependence on fossil fuels [21]. The view and awareness on the environmental issues have been reinforced with the latest UN IPCC reports on climate change concerns, published in 2021 [22].

In addition to the previously mentioned motivations, the EV field is also aroused by advances in the electricity industry. As an example, in the recent decade, electricity can also be produced with wind power on a large scale. In Sweden, there are a lot of offshore wind farms to generate electricity, a notable example is Lillgrund Wind Farm, in Øresund, near the Danish border [23, 24]. This simultaneous progress in both areas has been

FIGURE 5.4 Growth in the electricity sector stimulates the progress of vehicle electrification.

kimson/Shutterstock.com.

helpful in advancing the goal of reducing pollution through vehicle electrification (Figure 5.4).

In the UN SDGs, one of the themes is called Climate Action, which drives efforts to generate "carbon emission reduction." These discussions evoke electrification not only in the automotive sector but also in the whole mobility industry. Incentives from governments all around the world to support electrification are another factor to be considered, with notable examples of the US and Norway [25, 26]. It is therefore evident from the discussions in this section that vehicle electrification is motivated by various factors such as economic, societal, and technological benefits.

Despite the commonly heard mantra "future is electric," as mentioned in the previous section, it is still important to educate the general audience on the overview of the EV concept [27, 28].

What makes this book distinct from other related books is that the author tries to paint a big picture of future mobility from every aspect, which includes "autonomous driving" as the intelligence, "vehicle connectivity" as the method of communication, "vehicle electrification" as the basis of the powertrain, and the method on how these new vehicles are going to be employed in "shared mobility." As we have examined the first two elements in the ACES acronym in the previous chapters, in this chapter we will discuss vehicle electrification.

Accordingly, a technical overview of different types of EVs will be discussed in the next sections. This chapter aims to provide the general audience with a complete perspective on the concept of vehicle electrification and its role in enabling true ACES mobility.

5.3. Back-End Overview

Perhaps and arguably, one of the biggest stimulating factors of ACES, particularly EV progress, is the commitment agreed upon by most major carmakers at the 2021 UN Climate Change Conference (COP26) to produce zero-emission vehicles and phase out the sales of ICE vehicles by 2040 [29]. Although some major carmakers have yet to commit, most of those who committed show the potential of the EV field in contributing to this COP26 vision.

In this chapter, similar to Chapters 3 and 4, the reviews made are not aimed to provide exhaustive technical knowledge to the reader. Instead, because the book is targeting a general audience, we will look into different types of EVs and their brief summary in each subsection.

5.3.1. Hybrid Electric Vehicles

As with the cases of other disruptive technologies, the transformations do not happen overnight. When smartphone uprisings happened, they did not straightaway impact most of the world's population. Instead, before the adoption of smartphones like the iPhone or Samsung Galaxy series, most people were first introduced and familiarized with 3G phones with cameras by such classic examples as the Nokia XpressMusic Series [30]. Furthermore, technological shifts usually require a lot of support from infrastructures as well as regulations. That is also the reason why it took time for disruptive technologies like smartphones to gain traction in the market during the initial phase. When the external factors are readied for the public adoption of an innovation, market and business traction are also gained. The situation is comparable to what happened with Hybrid Electric Vehicles (HEVs), one of the earliest widely adopted usage of EV.

An HEV sees the cooperation between the "conventional" fuel-based engines and an electricity-powered motor to power the vehicle. On this platform, the ICE is run on natural fuel such as diesel or gasoline while the vehicle is also equipped with a battery for electrification purposes. One of the advantages of HEVs compared to conventional ICE vehicles is the reduced emissions from the vehicle. This is achieved by integrating the electrification elements as one of the power sources, alternatingly with the conventional vehicle. Accordingly, lower operating costs for the private users are also achieved (by integrating electricity as one of the energy sources, reducing fuel costs) [31].

In most cases, HEVs do not require an external mode of battery charging as the charging strategy operates on the vehicle platform itself. Using some decision-making strategy, the transition between ICE and electric charge transmission is determined [32]. Much work has been done to improve HEV power management to provide the best benefit to the users, both financially and from a ride comfort perspective. Several works have integrated machine-learning-based and other algorithms to improve the HEV power decision-making strategy between the ICE and electric motors (EMs) [32, 33].

Power management between ICE and EM in HEV also requires the consideration of different driving conditions. For example, a vehicle that is cruising along the highway at uniform speed can be using a battery as the sole power source (upon its availability), while the vehicle which is speeding along the highway might be using ICE or both ICE and EV, i.e., the battery can also be charged by the actions of the vehicle driver, such as braking. This technology is called regenerative braking [34–36]. Figure 5.5 illustrates the considerations required as input to the decision-making between the ICE and EM as power sources for HEVs.

FIGURE 5.5 Requirements for power management and decision-making formulations between ICE and EM as the main energy source for the HEV [37, 38].

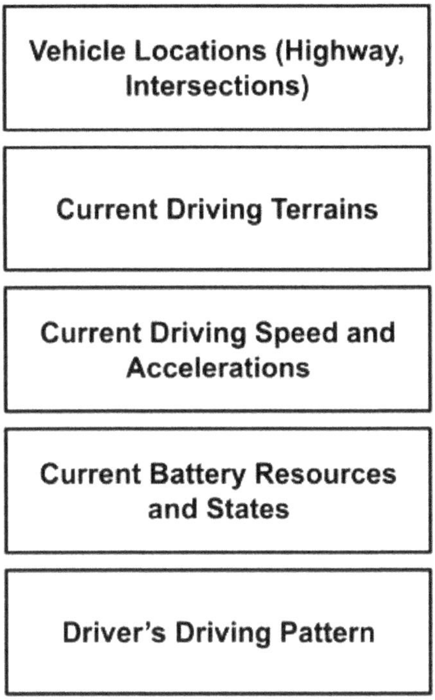

___

| Vehicle Locations (Highway, Intersections) |
| Current Driving Terrains |
| Current Driving Speed and Accelerations |
| Current Battery Resources and States |
| Driver's Driving Pattern |

© SAE International.

HEVs have successfully helped introduce the EV technology to early audiences. It is impossible for the market to embrace EV technology if the carmakers choose to bring full Battery Electric Vehicles (BEVs) right away as this will cause range anxiety [39]. With the transition from ICE to HEVs, the acceptance of EV technology has been more manageable compared to the immediate transition from ICE to BEVs.

HEVs also helped reduce emissions as well as fuel consumption while driving, for example, when the vehicle is at traffic stops [40, 41]. These economic benefits have succeeded in attracting a lot of sales in HEVs.

Not requiring external charging in most cases, also helped the adoption of HEVs in developing countries. In addition to the mentioned discussions, it is interesting to note another term that pertains to HEV topics, i.e., Mild HEV. The said vehicle allows the use of limited electric power for noncritical applications such as winter seat heating systems [42].

5.3.2. Plug-In Hybrid Electric Vehicles

To largely increase the acceptance of EV technology and enable mass adoption by the general public, a longer range of electric driving by EVs is required. Compared to HEVs, the main distinction of Plug-In Hybrid Electric Vehicles (PHEVs) is the availability of an external charging port on the vehicle that can be used for the charging grid. In addition, PHEVs also come with a larger battery pack and a more powerful EM compared to HEVs, allowing more EM-based driving by PHEVs [42].

This then allowed for the vehicle to travel for an additional range on electric mode (according to the article "BEV, PHEV or HEV: The Differences Affect the Architecture" [42], it can reach 30–50 km). Considering the vehicle is used for daily commute purposes, this will bring huge economic benefits to the user. Compared to the nominal HEV, Redelbach et al. [43] claimed that PHEVs show greater reductions in fuel consumption because of the extended all-electric driving range. However, this depends on the intervals between two battery charging sessions.

In addition to being able to charge with external sources, the PHEV battery, similar to HEV, can also be charged with regenerative braking (see the previous section "Hybrid Electric Vehicles"). Both HEVs and PHEVs can increase fuel economy savings by incorporating the start-stop technology [44]. The extended electric driving range makes PHEVs a very attractive package in the market, with vehicle deliveries reaching 2,264,400 globally in 2019, an increase of 9% from 2018 [45].

Demands for PHEVs are also attributed to the need for a more reliable, greener vehicle which includes lower emissions and quieter drives (refer to Chapter 2). According to Pencheva et al. [46], one of the methods to reduce urban noise pollution is the optimization of transportation modes and utilization of EVs. As the vehicle platform transitions to an all-electric mode, PHEVs

FIGURE 5.6 PHEVs allow an extended range of electric driving compared to HEVs because of the possibility of external charging.

Monkey Business Images/Shutterstock.com.

are a great innovation to motivate and encourage the use of EVs. As a reminder, it is important to note that the descriptions of PHEV in this subsection are the stereotype of the technology, and the actual PHEV quality may vary between car models of different brands (Figure 5.6).

5.3.3. Battery Electric Vehicles

Transportation electrification does not only happen in road vehicles. In fact, it is happening across the industry. For example, for urban mobility, in the past few years, we have seen a lot of companies promoting the use of shared electric scooters as a mode of transportation (Figure 5.7). It has proven to be a low-cost alternative to the already available transportation methods. One of the companies providing this service is Bird using XiaoMi scooters [47, 48]. Electrification is also being reinforced by various government and public awareness movements as previously mentioned, for example, the Biden regime's consumer rebates for EV sales [25].

FIGURE 5.7   Electrification efforts are happening across mobility sectors.

FXQuadro/Shutterstock.com.

In the early stage of EV introduction into the mass market, one of the biggest barriers to adoption is range anxiety. Based on the works of Rauh et al. [49] and Noel et al. [50], the definition of range anxiety can be oversimplified as the fear some people have of driving or using an electric car because the vehicle may stop running as the battery ran out of power before reaching the next charging station. This is a valid concern considering that not many years ago the available charging stations for EVs were few. As battery technology advances, companies will start trying to address this major pain point that has hampered EV adoption. For example, IONITY, a consortium of different vehicle manufacturers headquartered in Germany is trying to create a charging station network, allowing long-distance travels for EVs in Europe. Among the notable participants in the consortium are Daimler AG, BMW Group, and Volkswagen Group, among many others [51, 52]. Apart from improving the BEV battery, there are also several other alternative efforts introduced to improve the range anxiety concern. For example, to compensate the absence of charging grids, BEVs can also be driven with a range-extender trailer (i.e., portable charging spots) [53]. Subsequently, this has prompted the growth of BEVs.

BEVs can be defined as a vehicle that operates completely in EV mode. Compared to the PHEV and HEV, it does not have any ICE on its platform, therefore allowing a higher quantity of emission reductions caused by ICE-based vehicles. Technically, BEVs provide the desired vehicle motion torque by directly converting the electricity without the aid of an ICE. This allows for better vehicle accelerations and improves the driving experience [54, 55]. According to Huang et al. [56], BEVs allow an easier driving experience. This will then allow for newer drivers to learn better the vehicle operation, potentially improving the driving quality. As EV distance ranges depend on external factors such as driving style as well as weather [57], a lot of companies are trying to address this issue to improve the battery quality of BEVs.

The high number of BEV-related start-ups emerging with many investments in recent years shows the mass potential of EVs in the future. Furthermore, the increase in demand for in-vehicle infotainment by the market with BEVs facilitating the convenience of this feature has resulted in BEV sales continuously gaining traction. With improved electric driving range provided in the vehicle, BEVs potentially transform the concept of vehicle into a lifestyle, instead of a mere transportation platform. For example, MUJI, a Japanese lifestyle-retail company, and Sensible 4, a Finnish AD company, have exhibited the potential of ACES as a new lifestyle platform where it can be used not only for transportation but also as library and grocery stores [58]. Advances in BEV will be able to translate this vision into reality, where BEV will be an integral part of ACES.

5.3.4. Fuel Cell Electric Vehicle

Another type of EV, although less well known, are Fuel Cell Electric Vehicles (FCEVs). The onboard EMs are powered using the fusion of hydrogen and oxygen to generate electricity and water. A subsequent process then permits the hydrogen molecules to be divided into hydrogen ions and electrons. The electrons are then utilized

FIGURE 5.8 FCEVs.

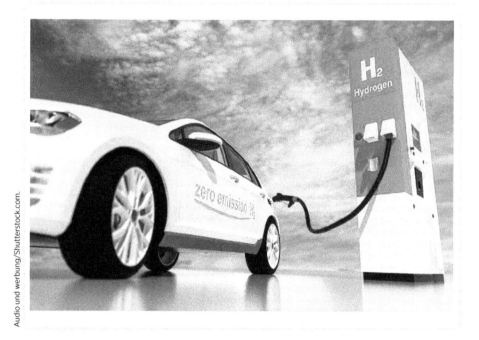

for the generation of electricity, which is then used for the FCEV driving. FCEVs also come with a battery which can be charged with the extra energy produced by the fuel cell. More in-depth detail on the topic of FCEVs can be found in the work by Hardman and Tal [59] (Figure 5.8).

Compared to BEVs, the FCEV is equipped with a hydrogen fuel tank and a fuel cell stack on its platform. According to the US Department of Energy website, a fuel cell stack is a membrane electrode assembly that generates electricity based on the utilization of hydrogen and oxygen. The FCEV also has a fuel filler on its platform to allow for refilling. Hydrogen is stored onboard the FCEV in the hydrogen fuel tank [60].

Similar to other types of EVs, the FCEV is equipped with power management and decision-making strategy. The power distributions are operated by a control unit [61]. For low-load driving, the power could be sourced from the fuel cell stack. In this scenario, regenerative braking is also used to charge the battery. However, for high-load driving such as highway or uphill driving, the FCEV might employ both battery and fuel cell stack components while simultaneously charging the battery from the regenerated braking. More discussions on this related topic can be found in works by Gomez et al. [62], and Salisbury [63].

The growth in the FCEV sector has led to an accumulation of FCEV-related jobs in the main job portals such as Indeed and Glassdoor [64, 65]. For example, a start-up in Singapore called Spectronik is focusing on developing hydrogen fuel-cell system technology with the aim to deliver autonomous FCEVs in the future [66].

However, despite the discussions, FCEV is relatively still not popular compared to its peers, such as HEV, PHEV, and BEV. The reasons include the small number of hydrogen fueling stations and related infrastructures, the safety challenges, and fewer media recognition [67].

5.3.5. Solar EVs

Despite not being quite popular, some companies utilize electricity as the main power source for their vehicles. Among the notable companies of Solar-based EVs (SEVs), or solar cars, are Sono Motors GmbH (based in Germany) and Lightyear (based in the Netherlands). Among their SEV models are Sono Motors Sion and Lightyear One [68, 69].

As the name suggests, SEVs are equipped with solar cells and power themselves using sunlight. With the considerable amounts of investments received by companies such as Lightyear to develop SEV (more than 100M USD as of February 2022), the discussion on SEVs is not something to be dismissed [70].

However, the challenges to the wide adoption of SEVs include the expensive price of solar cars. However, despite this, the author of the book assumes SEVs might have the potential for use in tropical countries (with the abundance of sunlight) such as the golf-buggy vehicle in holiday resorts as well as cargo mover in the airport area.

5.4. Vehicle Electrification Roles in Enabling True ACES Mobility

In the previous section, the author has provided an overview of different EV classifications. With regard to the role of vehicle electrification in the big picture of ACES mobility, EVs will play an important role in ACES mobility as reliable vehicle connectivity and safe AD feature of the AV require good computing platforms and sensors with stable electricity. For consistent safety of the AV during AD navigation, the sensors should always be working regardless of the weather conditions. These require good heating conditions during the winter season for the AV. Therefore, EVs will also help this purpose. Furthermore, since the aim of shared mobility is to enable emissions reduction, having EVs as part of the ACES will expedite this goal. Interested readers are invited to read more on this topic in the article by Kang et al. [71].

5.5. Vision and Current State of the EV Industry

EVs are one of the most "commercialized" future mobility topics. In many developed countries, particularly the Nordic region, EV sales are growing in addition to the active efforts to install the required infrastructures for EVs in the cities in the said region [72, 73].

The automotive industry is also seeing significant investments in EV-related companies such as REE Automotive, Arrival, and Canoo. In recent years, tech giants such as Amazon have also invested a significant amount of money into the automotive field. For example, RIVIAN, an EV company, has been working closely with Amazon to output their electric delivery van [74, 75].

To ensure the sustainability of EVs, the power source for the electrified platforms should also come from "green" and "sustainable" sources. Therefore, efforts should be made to produce more clean energy in different countries. More details on this topic will be discussed in Chapters 7, 8, and 9.

5.6. Summary

In this chapter, readers are provided with a comprehensive overview of vehicle electrification. The author believes the content of this chapter is sufficient to guide general readers from various disciplines with a better understanding of the concept and current state of the EV industry. Interested readers may refer to cited references for further details. The next chapter will provide an overview of shared mobility, which is the last puzzle of ACES mobility in this book.

References

1. Rațiu, S., "The History of the Internal Combustion Engine," *Annals of the Faculty of Engineering Hunedoara* 1, no. 3 (2003): 145-148.

2. Taylor, A.M.K.P., "Science Review of Internal Combustion Engines," *Energy Policy* 36, no. 12 (2008): 4657-4667.

3. Ahlbrandt, T.S. and Klett, T.R., "Comparison of Methods Used to Estimate Conventional Undiscovered Petroleum Resources: World Examples," *Natural Resources Research* 14, no. 3 (2005): 187-210.

4. Larminie, J. and Lowry, J., *Electric Vehicle Technology Explained* (Hoboken, NJ: John Wiley & Sons, 2012)

5. Situ, L., "Electric Vehicle Development: The Past, Present & Future," in *2009 3rd International Conference on Power Electronics Systems and Applications (PESA)*, Hong Kong, 1-3, IEEE, May 2009.

6. DeCotis, P.A., "The Rise of Transportation Electrification," *Natural Gas & Electricity* 36, no. 6 (2020): 9-15.

7. Cornell, B., Cornell, S., and Cornell, A., "Valuing the Automotive Industry," 2021.

8. Kumar, R.R. and Alok, K., "Adoption of Electric Vehicle: A Literature Review and Prospects for Sustainability," *Journal of Cleaner Production* 253 (2020): 119911.

9. Ithnin, A.M., Yahya, W.J., Ahmad, M.A., Ramlan, N.A. et al., "Emulsifier-Free Water-in-Diesel Emulsion Fuel: Its Stability Behaviour, Engine Performance and Exhaust Emission," *Fuel* 215 (2018): 454-452.

10. Taylor, B., "Fortum to Build EV Battery Recycling Plant," Recycling Today, accessed August 2022, https://www.recyclingtoday.com/article/fortum-finland-ev-lithium-ion-battery-recycling-investment/

11. Helen Ventures, "Four Million Euros to Be Invested in the Electric Vehicle Charging Company Virta Ltd,", accessed August 2022, https://www.helenventures.fi/stories/four-million-euros-to-be-invested-in-the-electric-vehicle-charging-company-virta-ltd

12. Johnner Olin, A. and Penser, T., "Light Electric Freight Vehicle Concept: A Pedal Assist Trike Designed for Urban Last Mile Deliveries," 2021.

13. Tarkowski, M., "On the Emergence of Sociotechnical Regimes of Electric Urban Water Transit Systems," *Energies* 14, no. 19 (2021): 6111.

14. Arrival Website, accessed August 2022, https://arrival.com/world/en/

15. Youtube, "Will Ferrell Super Bowl Ad—General Motors [2021]," accessed August 2022, https://youtu.be/mdsPvbSpB2Y

16. Youtube, "GM Super Bowl Clip," accessed February 2022, https://youtu.be/OSEtu49xkqY

17. Bloomberg Europé Edition, "BMW's Super Bowl Pick for Selling EVs: Schwarzenegger," accessed February 2022, https://www.bloomberg.com/news/articles/2022-02-09/schwarzenegger-plays-electrified-zeus-in-bmw-s-new-ev-ads

18. Pevec, D., Babic, J., Carvalho, A., Ghiassi-Farrokhfal, Y. et al., "Electric Vehicle Range Anxiety: An Obstacle for the Personal Transportation (r) Evolution?," in *2019 4th International Conference on Smart and Sustainable Technologies (SPLITECH)*, Split/Bol, Croatia, 1-8, IEEE, 2019.

19. Cleary, K., "Electrification 101," Resources for the Future, accessed August 2022, https://www.rff.org/publications/explainers/electrification-101/

20. Hagman, J., Ritzén, S., Stier, J.J., and Susilo, Y., "Total Cost of Ownership and Its Potential Implications for Battery Electric Vehicle Diffusion," *Research in Transportation Business & Management* 18 (2016): 11-17.

21. Adnan, N., Nordin, S.M., Rahman, I., and Amini, M.H., "A Market Modeling Review Study on Predicting Malaysian Consumer Behavior towards Widespread Adoption of PHEV/EV," *Environmental Science and Pollution Research* 24, no. 22 (2017): 17955-17975.

22. IPCC, "Climate Change 2021: The Physical Science Basis," in Masson-Delmotte, V., Zhai, P., Pirani, A., Connors, S.L. et al. (eds.), *Contribution of Working Group I to the Sixth Assessment Report of the Intergovernmental Panel on Climate Change* (Cambridge: Cambridge University Press, 2021), 2.

23. Sweden.Se, "Energy Use in Sweden," accessed August 2022, https://sweden.se/climate/sustainability/energy-use-in-sweden

24. Vattenfall Powerplants, "Lillgrund—The Largest Offshore Wind Farm in Sweden," accessed July 2022, https://powerplants.vattenfall.com/lillgrund

25. Reuters, " Biden Plan Calls for $100 Billion in New EV Consumer Rebates: Email," accessed July 2022, https://www.reuters.com/article/us-usa-autos-idUSKBN2BU3BA

26. Norsk elbilforening, " Norwegian EV Policy," accessed February 2022, https://elbil.no/english/norwegian-ev-policy

27. Lashof, D. and Gorguinpour, C., "Memo to Carmakers: The Future Is Electric," 2018, https://www.wri.org/insights/memo-carmakers-future-electric.

28. Hannisdahl, O.H., Malvik, H.V., and Wensaas, G.B., "The Future Is Electric! The EV Revolution in Norway—Explanations and Lessons Learned," in *2013 World Electric Vehicle Symposium and Exhibition (EVS27)*, Barcelona, Spain, 1-13, IEEE, November 2013.

29. Mountford, H., Waskow, D., Gonzalez, L., Gajjar, C. et al., "COP26: Key Outcomes from the UN Climate Talks in Glasgow," 2021.

30. Alibage, A. and Weber, C., "Nokia Phones: From a Total Success to a Total Fiasco: A Study on Why Nokia Eventually Failed to Connect People, and an Analysis of What the New Home of Nokia Phones Must Do to Succeed," in *2018 Portland International Conference on Management of Engineering and Technology (PICMET)*, Honolulu, HI, 1-15, IEEE, 2018.

31. Alternative Fuels Data Center, "Hybrid Electric Vehicles," US Department of Energy, accessed August 2022, https://afdc.energy.gov/vehicles/electric_basics_hev.html

32. Zhao, P., Wang, Y., Chang, N., Zhu, Q. et al., "A Deep Reinforcement Learning Framework for Optimizing Fuel Economy of Hybrid Electric Vehicles," in *2018 23rd Asia and South Pacific Design Automation Conference (ASP-DAC)*, San Francisco, CA, 196-202, IEEE, 2018.

33. Lin, X., Wang, Y., Bogdan, P., Chang, N. et al., "Optimizing Fuel Economy of Hybrid Electric Vehicles Using a Markov Decision Process Model," in *2015 IEEE Intelligent Vehicles Symposium (IV)*, Seoul, South Korea, 718-723, IEEE, 2015.

34. Panday, A. and Bansal, H.O., "A Review of Optimal Energy Management Strategies for Hybrid Electric Vehicle," *International Journal of Vehicular Technology* 2014 (2014): 1-19.

35. Ali, A.M. and Söffker, D., "Towards Optimal Power Management of Hybrid Electric Vehicles in Real-Time: A Review on Methods, Challenges, and State-of-the-Art Solutions," *Energies* 11, no. 3 (2018): 476.

36. Wei, Z., Xu, J., and Halim, D., "HEV Power Management Control Strategy for Urban Driving," *Applied Energy* 194 (2017): 705-714.

37. Wang, S. and Lin, X., "Eco-Driving Control of Connected and Automated Hybrid Vehicles in Mixed Driving Scenarios," *Applied Energy* 271 (2020): 115233.

38. Cairano, D., Stefano, D.B., Bemporad, A., and Kolmanovsky, I.V., "Stochastic MPC with Learning for Driver-Predictive Vehicle Control and Its Application to HEV Energy Management," *IEEE Transactions on Control Systems Technology* 22, no. 3 (2013): 1018-1031.

39. Melliger, M.A., van Vliet, O.P., and Liimatainen, H., "Anxiety vs Reality-Sufficiency of Battery Electric Vehicle Range in Switzerland and Finland," *Transportation Research Part D: Transport and Environment* 65 (2018): 101-115.

40. Tang, X., Duan, Z., Hu, X., Pu, H. et al., "Improving Ride Comfort and Fuel Economy of Connected Hybrid Electric Vehicles Based on Traffic Signals and Real Road Information," *IEEE Transactions on Vehicular Technology* 70, no. 4 (2021): 3101-3112.

41. Rahmani, D. and Loureiro, M.L., "Assessing Drivers' Preferences for Hybrid Electric Vehicles (HEV) in Spain," *Research in Transportation Economics* 73 (2019): 89-97.

42. APTIV, "BEV, PHEV or HEV: The Differences Affect the Architecture," accessed August 2022, https://www.aptiv.com/en/insights/article/bev-phev-or-hev-the-differences-affect-the-architecture

43. Redelbach, M., Friedrich, H.E., Le Berr, F., Rousseau, A. et al., "Comparison of Energy Consumption and Costs of Different HEVs and PHEVs in European and American Context," 2012.

44. Rousseau, A. and Ding, Y., "Impact of Worldwide Test Procedures on Advanced Technology Fuel Efficiency Benefits," in *2013 World Electric Vehicle Symposium and Exhibition (EVS27)*, Barcelona, Spain, 1-8, IEEE, November 2013.

45. EV Volumes, "Global BEV & PHEV Sales for 2019," accessed February 2022, https://www.ev-volumes.com/news/global-bev-phev-sales-for-2019

46. Pencheva, V., Beloev, I., Asenov, A., and Topchu, D., "A Study on the Noise Pollution from Traffic Flows," in *Scientific Forum Challenges in Engineering and Information Science*, Sinaia, Romania, 2015.

47. Bird's Website, accessed February 2022, https://www.bird.co/

48. Xiaomi, "Mi Electric Scooter Pro," accessed August 2022, https://www.mi.com/global/mi-electric-scooter-pro

49. Rauh, N., Franke, T., and Krems, J.F., "Understanding the Impact of Electric Vehicle Driving Experience on Range Anxiety," *Human Factors* 57, no. 1 (2015): 177-187.

50. Noel, L., de Rubens, G.Z., Sovacool, B.K., and Kester, J., "Fear and Loathing of Electric Vehicles: The Reactionary Rhetoric of Range Anxiety," *Energy Research & Social Science* 48 (2019): 96-107.

51. IONITY's Website, "Enabling Electric Travel for Everyone," accessed August 2022, https://ionity.eu/

52. Dižo, J., Blatnický, M., Semenov, S., Mikhailov, E. et al., "Electric and Plug-In Hybrid Vehicles and Their Infrastructure in a Particular European Region," *Transportation Research Procedia* 55 (2021): 629-636.

53. InsideEVs, "Nomadic Power Presents Range Extender Trailer (w/video)," accessed August 2022, https://insideevs.com/news/322567/nomadic-power-presents-range-extender-trailer-w-video

54. Alternative Fuels Data Center, "All-Electric Vehicles," US Department of Energy, accessed August 2022, https://afdc.energy.gov/vehicles/electric_basics_ev.html

55. Wallbox, "How Do EVs Compare to Gas Cars?," accessed August 2022, https://blog.wallbox.com/how-do-evs-compare-to-gas-cars

56. Huang, X., Lin, Y., Lim, M.K., Tseng, M.L. et al., "The Influence of Knowledge Management on Adoption Intention of Electric Vehicles: Perspective on Technological Knowledge," *Industrial Management & Data Systems* 121, no. 7 (2021): 1481-1495.

57. Donkers, A., Yang, D., and Viktorović, M., "Influence of Driving Style, Infrastructure, Weather and Traffic on Electric Vehicle Performance," *Transportation Research Part D: Transport and Environment* 88 (2020): 102569.

58. Youtube, "MUJI無印良品:GACHA Autonomous Shuttle Bus," MUJIglobal, accessed August 2022, https://youtu.be/W0oxz9KpiIY

59. Hardman, S. and Tal, G., "Who Are the Early Adopters of Fuel Cell Vehicles?" *International Journal of Hydrogen Energy* 43, no. 37 (2018): 17857-17866.

60. Alternative Fuels Data Center, "How Do Fuel Cell Electric Vehicles Work Using Hydrogen?," US Department of Energy, accessed August 2022, https://afdc.energy.gov/vehicles/how-do-fuel-cell-electric-cars-work

61. Mahmoudi, C., Flah, A., and Sbita, L., "An Overview of Electric Vehicle Concept and Power Management Strategies," in *2014 International Conference on Electrical Sciences and Technologies in Maghreb (CISTEM)*, Tunis, Tunisia, 1-8, IEEE, 2014.

62. Gomez, A., Sasmito, A.P., and Shamim, T., "Investigation of the Purging Effect on a Dead-End Anode PEM Fuel Cell-Powered Vehicle during Segments of a European Driving Cycle," *Energy Conversion and Management* 106 (2015): 951-957.

63. Salisbury, S., "Understanding Fuel Cell Plug-In Hybrid Electric Vehicle Use, Design, and Functionality," PhD dissertation, Colorado State University, 2014.

64. Indeed, "Fuel Cell Engineer," accessed August 2022, https://www.indeed.com/q-Fuel-Cell-Engineer-jobs.html

65. Glassdoor, "Fuel Cell Engineer," accessed August 2022, https://www.glassdoor.com/Job/fuel-cell-engineer-jobs-SRCH_KO0,18.htm

66. Spectronik's Website, "PROTIUM: Air-Cooled Fuel Cells and Accessories," accessed August 2022, https://www.spectronik.com/

67. Tanç, B., Arat, H.T., Baltacıoğlu, E., and Aydın, K., "Overview of the Next Quarter Century Vision of Hydrogen Fuel Cell Electric Vehicles," *International Journal of Hydrogen Energy* 44, no. 20 (2019): 10120-10128.

68. Sono's Website, "Solar Integration," accessed August 2022, https://sonomotors.com

69. Lightyear's Website, "Freedom to Move Anywhere," accessed August 2022, https://lightyear.one/lightyear-one

70. Crunchbase, "Lightyear," accessed August 2022, https://www.crunchbase.com/organization/lightyear

71. Kang, N., Feinberg, F.M., and Papalambros, P.Y., "Autonomous Electric Vehicle Sharing System Design," *Journal of Mechanical Design* 139, no. 1 (2017): 011402.

72. Kester, J., Sovacool, B.K., Noel, L., and de Rubens, G.Z., "Rethinking the Spatiality of Nordic Electric Vehicles and Their Popularity in Urban Environments: Moving Beyond the City?" *Journal of Transport Geography* 82 (2020): 102557.

73. Kester, J., Sovacool, B.K., de Rubens, G.Z., and Noel, L., "Novel or Normal? Electric Vehicles and the Dialectic Transition of Nordic Automobility," *Energy Research & Social Science* 69 (2020): 101642.

74. Garrido, C., "The Impact of Amazon Delivery Van Electrification," 2020.

75. Stinson, J., "Amazon Buys 100,000 Electric Vans from Rivian for Prime Deliveries," Transport Topics, 2019, https://trid.trb.org/view/1663737

6

Shared Mobility: Concise Overview

6.1. Background

We have examined autonomous, connected, and electric vehicle technicalities in the previous chapters. But the one entity that binds them all to form ACES as the future of transportation and automotive is "Shared Mobility."

Shared mobility connects the mentioned elements to form a future shared driverless electric mobility. Without the "sharing" features, the potentially massive disruptions brought by ACES will not transpire [1, 2]. Shared mobility not only transforms the automotive and mobility sectors from the technical aspect but also enables transformations in the concept of vehicle ownership, which will then reduce the number of private vehicle ownerships [3].

It also boosts the concepts of digitalized and personalized in-cabin experience, where the vehicle is expected as something no longer privately owned but a platform for mobility from point A to point B [4]. Several decades ago, this might be a strange idea— that is, sharing vehicles with other people—but not anymore with the blossoming of shared mobility.

In this chapter, the author will first highlight the motivation and justification behind the shared mobility expansion. And since shared mobility spurred a lot of terms that sound identical like "carsharing," "ridesharing," and "carpooling," among many others, the author will briefly review each well-known term and its current application. This will provide the general reader with a bird's-eye view to better understand these concepts. Also, additional reviews on the role of shared mobility in the ACES ecosystem are discussed.

6.2. Motivations behind Shared Mobility

Several developments and progress stimulate the development of shared mobility. Some of them are listed in this section.

6.2.1. Sharing Economy

The author has mentioned the rise of the sharing economy in some of the earlier chapters. The passage below denotes the discussion of sharing economy, as written in Chapter 2.

> According to the World Economic Forum, sharing economy means the sharing of underutilized assets, monetized or not, in ways that improve efficiency, sustainability, and community [5]. With the advances in Internet technology, the FIR, and digital awareness among world inhabitants, sharing economy has been a famous discussion for "future economy." Perhaps the most popular example, Airbnb Inc., has managed to indirectly disrupt the tourism ecosystem and hotel industry by allowing private landlords to lease their apartments to tourists. By cutting the middleman, Airbnb has facilitated the growth of sharing economy [5]. Similar to that, as mentioned, mobility platforms such as Uber, Gojek, Grab, and Bolt (previously Taxify) have boosted sharing economy applications in different regions [6, 7]. This has then supported the growth of shared mobility as part of ACES.

The quote above is intentionally duplicated here for better understanding by readers who have only read this chapter. Inquisitive readers are invited to purchase and read the remaining chapters to get a better overview of ACES mobility. From the excerpt, we recognize the growth of the sharing economy and how

FIGURE 6.1 The growth of sharing economy influencing the shared mobility progress.

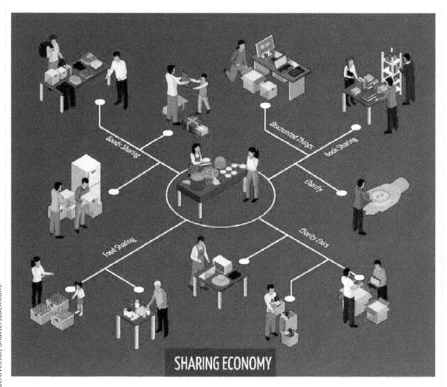

Macrovector/Shutterstock.com.

it will influence the technologies within the FIR umbrella (Figure 6.1). Even subscription-based entertainment service companies like Netflix and Spotify have now introduced the "family plan" in their pricing plan, which indicates the market demand for a "sharing economy" with the said account sharing [8]. These developments and progress signal the changes in behavior among the users. Consequently, sharing economy has now influenced a lot of industries, including automotive and mobility. For example, we are seeing the growth of companies like Uber, Lyft, Grab, Gojek, Bolt, and Yango which basically is the indirect outcome of the sharing economy [9]. The growth will be materialized more soon when it starts to link with other emerging automotive technology such as ACES ecosystems.

6.2.2. Uberization

The maturation of sharing economy disrupted the mobility sector, for example, the taxi endeavor. Uber, as one of the first giants who became successful from the movement, managed to alter the behavior of city dwellers that they now prefer services like Uber ridehailing instead of calling for taxis [10].

Uber and its peers have all received huge investments globally, and their success especially concocted the term *uberization*. From the Cambridge Online Dictionary, the said word refers to "the act or process of changing the market for a service by introducing a different way of buying or using it, especially using mobile technology" [11].

With the growth of both sharing economy and mobile technology, Mobility-as-a-Service (MaaS) maintains to instigate the evolution of shared mobility. This has then catalyzed more ambitious collaboration by similar companies. For example, Shotl, an on-demand shuttle provider company in Spain, has collaborated with Finnish AD company, Sensible 4, for their pilot project of shared driverless mobility [12]. The idea is to boost on-demand shared driverless mobility by collaborating with the Sensible 4 AD shuttle bus. This then will allow for better last-mile solutions for city transportation. This has managed to catalyze more discussions in the world of ACES technology, which eventually will disrupt the whole automotive ecosystem.

6.2.3. X-as-a-Service

Sharing economy is heavily dependent on mobile technology and the Internet. For an innovation or product to be shared with other people, a user should have Internet connectivity, and this should be enabled by a middle-layer software that is equipped with technologies such as cloud and data sharing. This has helped to grow many XaaS applications.

Accordingly, we are seeing more and more sharing economy activities which integrate the XaaS philosophy in their business model, for example, cloud storage which can be shared with different account users [13]. The growth of XaaS also has helped paint a clearer picture for automotive giants to transform their companies. More and more automotive companies and mobility start-ups have started talking about XaaS and sharing economy, and this has assisted to prompt the advancement of the shared mobility space.

6.3. Terminology and Definition: Clarification and Difference

Shared mobility is a rapidly growing field. And it is not only expanding in one country but developing globally in different continents, in different nations and languages. Consequently, much confusion arises concerning the terminology of the technology on this topic.

This subsequently might add challenges to the legalization and regulation improvement process. In fact, since the technologies are already in the market, the occasional use of words of the technologies is confusing. Therefore, the author of this book is convinced to clarify the distinction between the terminologies in the shared mobility area. This chapter, along with other chapters, will provide new readers with a clear overall picture of the ACES ecosystem.

In this section, we are going to review the popular terminologies related to shared mobility. The discussion will be based on several documents by SAE International, such as SAE J3163™ (Taxonomy of On-Demand and Shared Mobility: Ground, Aviation, and Marine) [14], the summary of SAE J3163™ [15], and a white-paper entitled "Definitions for Terms Related to Shared Mobility and Enabling Technologies" [16].

According to SAE J3163™ [14] and the summary of SAE J3163™ [15], among the key phrases for shared mobility are "shared mobility," "bike sharing," "carsharing," "microtransit," "ridesharing," "ridesourcing," and "scooter sharing." Consequently, we are going to review some of these jargons and other related terminologies in their own subsections. It is necessary to note that the discussions will be mostly on the terminology, and not the back-end technology of the words. The author believes it is a necessity to understand the terms, and the knowledge will allow each of the stakeholders in the automotive and mobility ecosystem to facilitate safer shared mobility delivery, which is reliable and fulfills customer expectations.

6.3.1. Shared Mobility

According to SAE J3163™ [14] and the summary of SAE J3163™ [15], shared mobility is the shared use of any vehicular platform for travel mode including, but not limited to, vehicles, motorcycles, and bicycles. Shared mobility provides users access to transportation and mobility for a limited short period of time upon request. The on-demand feature of shared mobility refers to *the ability to reserve or dispatch a service upon request by users* [14, 15]. Shared mobility is crucial for ACES mobility as it will see the on-demand robotaxi applications where the said platform will be shared with different users, enabled by the AV, CV, and EV technologies.

6.3.2. Carsharing

Carsharing refers to the provision of vehicle access to the registered members of an organization. Carsharing also provides the maintenance fleet by the organization, where the vehicles can be cars or trucks. The service requires payments of fees by members to the organization, and this can be monthly or per use. The benefit of carsharing is that it allows users to use vehicles without actually owning the vehicle, therefore reducing the hassle of vehicle maintenance [14, 15]. Companies like Lynk & Co are utilizing the sharing philosophy in their business model [17].

6.3.3. Carpooling

Long before the sharing economy and mobile technology gained traction in the market, the essence of shared mobility is already in the form of carpooling [18] (Figure 6.2). Carpooling refers to the

FIGURE 6.2 Carpooling is one of the most well-known terms related to shared mobility.

sharing of vehicles by several people going to the same destination, usually with the same route. Carpooling has been adopted by people in major cities around the world to reduce costs for participating members. The regulatory bodies in cities also encourage carpooling as it reduces the number of vehicles in the cities during peak hours. The driving responsibility can also be shared among the carpool members in turns. According to SAE J3163™ [14] and the summary of SAE J3163™ [15], carpooling is listed in the same category as ridesharing.

6.3.4. Ridesharing

Similar to carpooling, it is a sharing of rides between passengers and drivers with similar routes to a defined destination. Please refer to the carpooling section and its cited references for details [14, 15, 16].

6.3.5. Ridehailing

SAE J3163™ [14] and the summary of SAE J3163™ [15] categorize ridehailing in the same category as Transportation Network Company (TNC) and ridesourcing. It is a mobility and transportation service provided by a company and digitally enabled by mobile applications or platforms which facilitate payments and financial transactions. Ridehailing, TNC, and ridesourcing themselves are not similar to ridesharing or carpooling as ridehailing is a for-hire vehicle service whereas carpooling is more of sharing between drivers and travelers. More details can be found in SAE J3163™ [14], the summary of SAE J3163™ [15], and "Definitions for Terms Related to Shared Mobility and Enabling Technologies" [16].

6.3.6. Ridesourcing

According to "Carpooling: Problems and Potentials" [18], ridesourcing is a booked service of on-demand mobility using the means of digital mobile platforms. It is similar to ridehailing and TNC where a for-hire vehicle service is provided to the passengers. Details can be found in SAE J3163™ [14] and the summary of SAE J3163™ [15], which is published by the SAE Shared and Digital Mobility Committee [19], and in the previous subsection on Ridehailing.

6.3.7. Micromobility

Micromobility is a mode of transportation or mobility using small and lightweight types of lower-speed vehicles such as bicycles and scooters, among many others (Figure 6.3). Shared micromobility, on the other hand, is the shared use of micromobility platforms such as bicycles and mopeds over a particular period on an on-demand basis [14, 15, 16]. Further readings can also be found in SAE J3194, Taxonomy and Classification of Powered Micromobility Vehicles [20].

FIGURE 6.3 Micromobility has also grown in recent decades, particularly in urban areas.

Ground Picture/Shutterstock.com.

6.3.8. Paratransit

Paratransit refers to the provision of transportation and mobility services for individualized transportation needs. It is usually provided to individuals with special needs such as wheelchairs, among many others, as a solution for their mobility needs [21].

6.3.9. Microtransit

According to SAE J3163™ [14] and the summary of SAE J3163™ [15], microtransit refers to a transportation service that uses multi-passenger shuttles or vans to supply on-demand or fixed-schedule services. It is usually operated by public or private entities, and the routing can be dynamic or fixed.

6.3.10. Other Types of Shared Mobility

Apart from the mentioned terminologies above, other types of shared mobility-related terms will not be explored in this section but might be interesting for certain readers. These include bike sharing, scooter sharing, ridesplitting, and ground last-mile delivery services. More details can be found in SAE J3163™ [14], the summary of SAE J3163™ [15], "Definitions for Terms Related to Shared Mobility and Enabling Technologies" [16], and the works by Chen et al. [22] and Sonneberg et al. [23].

6.4. Differences between "Traditional Taxi" and "Shared Mobility"

From the previous section, we have identified the relevant terminologies related to shared mobility. But one might still be questioning what is the difference between a traditional conventional taxi and shared mobility services. According to reports and articles, the difference lies in the context of *ride-fare, vehicle ownerships*, the *availability of dynamic routing instead of fixed routing, service quality*, and the *driver or operator's experience*. Details can be found in the report by Correa et al. [24], the article "Difference between Taxi and Uber" [25], and the review by Cramer and Kruger [26].

Since shared mobility is a growing field, the "consumer-facing" service design philosophy adopted by the technology companies operating the shared mobility will promise more customer experience improvements. Therefore, we are seeing more "traditional car rental" and taxi companies trying to transform their business model to fit the future mobility narrative [27, 28].

6.5. Customer-and-User-Facing Technology Development for Shared Mobility

According to SAE J3163™ [14] and the summary of SAE J3163™ [15], apart from the popular nomenclatures that we have discussed in Section 1.2, there are also other terms that might be interesting for readers. For example, for business models development for future shared mobility, the terminologies include:

- Business-to-Business (B2B)
- Business-to-Consumer (B2C)
- Business-to-Government (B2G)
- Peer-to-Peer (P2P)
- Fractional Ownership
- Subscription Service

As can be seen, shared mobility is highly disruptive in nature, requiring automakers to collaborate, partner, and transform their business and operating models to remain relevant. The potential multi-segment business also requires discussions with different stakeholders in the ecosystem such as government institutions, nonautomotive companies, and public bodies. This is necessary to ensure a reliable service is developed, meeting the expectation of customers and providing them with the desired value. Sometimes, a simple innovation can be attractive enough to allow customer acquisition and retention. Companies like MaaS Global Oy in Helsinki managed to attract a lot of users by introducing a one-stop solution mobility application with their mobile app Whim [29].

Similarly, for delivery purposes, companies like Airmee start to gain traction in the market by solving the "waiting dilemma" pain point for parcel receivers [30]. These are the examples that demon-strate that shared mobility is really dependent on the customer requirements, similar to other software-based businesses. Therefore, R&D should not only focus on the back-end technologies but also on developing a solution that attracts and retains customers.

6.6. Roles of Shared Mobility in the ACES Ecosystem

The next questions from readers after reading the previous sections might be "So how can all these autonomous, connected, electric, and shared mobility be fused together, and what is the role of the shared mobility in this picture?"

The answer will be that shared mobility will be the element closest to the end users compared to the other entities in the ACES. For example, shared mobility will change the user behavior toward mobility. Shared mobility companies will also produce the required on-demand fleet mobile applications that will be employed by the users. All these require a customer-focus mindset in the automotive and mobility industries (Figure 6.4).

One of the best examples of shared driverless mobility is illustrated in a video by MUJI (Japanese fashion brand) and Sensible 4 (Finnish AD company) [31]. The video shows that shared mobility as part of the ACES ecosystem will not only bring driverless technology to the masses but will also transform the definition of mobility to the public, where the vehicle can not only be used for

FIGURE 6.4   Shared mobility will play a big role in the ACES ecosystem.

Chesky/Shutterstock.com.

transporting but also as a mobile market, mobile library, and possible mobile clinics. With companies like Ree, Zoox, Sensible 4, and Navya building autonomous shuttles for future shared mobility use cases with distinctive designs, they give a glimpse of the potential impact that shared mobility will bring to human society.

6.7. Summary

In this chapter, the author provided an overview of the terminologies for shared mobility and their relation with the whole ACES ecosystem. Clearly, shared mobility will play a major role in the future of transport. In fact, we are witnessing more automotive companies and organizations develop their output from the perspective of "mobility" instead of "conventional automotive." This requires cross-collaboration and a lot of engagements with different stakeholders in the growing ACES ecosystem (Figure 6.5). With the arrival of emerging discussions such as metaverse and personalized in-cabin infotainment, the demands for more

FIGURE 6.5 Integrating future shared mobility into the current mobility and transportation ecosystem requires a lot of collaboration and discussions between stakeholders.

Blue Planet Studio/Shutterstock.com.

customer-facing product development and operational organizations must be accomplished for the future survival of automotive companies. In conclusion, when it comes to shared mobility, the actual potential challenge that the industry will face is *"how to integrate and fuse the future shared mobility of ACES into the current mobility ecosystem."*

References

1. Hu, J.-W. and Creutzig, F., "A Systematic Review on Shared Mobility in China," *International Journal of Sustainable Transportation* 16 (2021): 374-389.

2. Machado, C.A.S., de Salles Hue, N.P.M., Berssaneti, F.T., and Quintanilha, J.A., "An Overview of Shared Mobility," *Sustainability* 10, no. 12 (2018): 4342.

3. Golbabaei, F., Yigitcanlar, T., and Bunker, J., "The Role of Shared Autonomous Vehicle Systems in Delivering Smart Urban Mobility: A Systematic Review of the Literature," *International Journal of Sustainable Transportation* 15, no. 10 (2021): 731-748.

4. Kuoch, S.-K., Nowakowski, C., Hottelart, K., Reilhac, P. et al., "Designing an Intuitive Driving Experience in a Digital World," *Preprints* (2018): 2018070629, doi:10.20944/preprints201807.0629.v1

5. Rinne, A., "What Exactly Is the Sharing Economy?" World Economic Forum, accessed August 2022, https://www.weforum.org/agenda/2017/12/when-is-sharing-not-really-sharing/

6. Zervas, G., Proserpio, D., and Byers, J.W., "The Rise of the Sharing Economy: Estimating the Impact of Airbnb on the Hotel Industry," *Journal of Marketing Research* 54, no. 5 (2017): 687-705.

7. Kurniawati, D.E. and Khoirina, R.Z., "Online-Based Transportation Business Competition Model of Gojek and Grab," *Advances in Social Science, Education and Humanities Research* 436 (2020): 1054-1057.

8. Obada-Obieh, B., Huang, Y., and Beznosov, K., "The Burden of Ending Online Account Sharing," in *Proceedings of the 2020 CHI Conference on Human Factors in Computing Systems*, Honolulu, HI, 1-13, 2020.

9. Khavarian-Garmsir, A.R., Sharifi, A., Abadi, H.H., and M., "The Social, Economic, and Environmental Impacts of Ridesourcing Services: A Literature Review," *Future Transportation* 1, no. 2 (2021): 268-289.

10. Nistal, P.D. and Regidor, J.R.F., "Comparative Study of Uber and Regular Taxi Service Characteristics," in *Proceedings of the 23rd Annual Conference of the Transportation Science Society of the Philippines*, Quezon City, Philippines, 2016, accessed August 13, 2017, http://ncts.upd.edu.ph/tssp/wp-content/upload/2016/08/Paronda-et-al.pdf

11. Cambridge Dictionary, "Uberization," accessed August 2022, https://dictionary.cambridge.org/dictionary/english/uberization

12. Shotl, "Shotl and Sensible 4 Adapt Its Autonomous Driving Pilot in Helsinki," accessed February 2022, https://shotl.com/news/shotl-and-sensible-4-adapt-its-autonomous-driving-pilot-in-helsinki-du

13. Alotaibi, S., Alomair, H., and Elhussein, M., "Comparing Performance of Commercial Cloud Storage Systems: The Case of Dropbox and One Drive," in *2019 International Conference on Computer and Information Sciences (ICCIS)*, Wuhan, China, 1-5, IEEE, April 2019.

14. SAE International, "Taxonomy of On-Demand and Shared Mobility: Ground, Aviation, and Marine," accessed February 2022, https://www.sae.org/standards/content/ja3163_202106/

15. SAE International, "Shared Mobility Taxonomy and Definitions in SAE J3163™," accessed February 2022, https://www.sae.org/binaries/content/assets/cm/content/topics/shared-mobility/summary-of-j3163.pdf

16. SAE International, "Definitions for Terms Related to Shared Mobility and Enabling Technologies WP-0010," accessed February 2022, https://www.sae.org/publications/technical-papers/content/wp-0010/

17. Lynk & Co, "Sharing," accessed February 2022, https://www.lynkco.com/en/sharing

18. Oppenheim, N., "Carpooling: Problems and Potentials," *Traffic Quarterly* 33, no. 2 (1979): 253-262.

19. SAE International, "Shared Mobility," accessed February 2022, https://www.sae.org/shared-mobility/

20. SAE International, "Taxonomy and Classification of Powered Micromobility Vehicles," accessed February 2022, https://www.sae.org/standards/content/j3194_201911

21. Regional Transit Service, "What Is Paratransit?," accessed February 2022, https://www.myrts.com/RTS-Access/What-is-Paratransit

22. Chen, X.M., Zahiri, M., and Zhang, S., "Understanding Ridesplitting Behavior of On-Demand Ride Services: An Ensemble Learning Approach," *Transportation Research Part C: Emerging Technologies* 76 (2017): 51-70.

23. Sonneberg, M.-O., Leyerer, M., Kleinschmidt, A., Knigge, F. et al., "Autonomous Unmanned Ground Vehicles for Urban Logistics: Optimization of Last Mile Delivery Operations," in *Proceedings of the 52nd Hawaii International Conference on System Sciences*, Maui, HI, 2019.

24. Correa, D., Xie, K., and Ozbay, K., "Exploring the Taxi and Uber Demand in New York City: An Empirical Analysis and Spatial Modeling," in *96th Annual Meeting of the Transportation Research Board*, Washington, DC, 2017.

25. Yadav, P., "Difference between Taxi and Uber," accessed February 2022, https://askanydifference.com/difference-between-taxi-and-uber

26. Cramer, J. and Krueger, A.B., "Disruptive Change in the Taxi Business: The Case of Uber," *American Economic Review* 106, no. 5 (2016): 177-182.

27. Korn, J., "Hertz to Buy up to 65,000 Electric Cars from Polestar," CNN, accessed April 2022, https://edition.cnn.com/2022/04/04/tech/polestar-hertz-purchase/index.html

28. NY Times, "Tesla Value Tops $1 Trillion after Hertz Orders 100,000 Cars," accessed August 2022, https://www.nytimes.com/2021/10/25/business/hertz-tesla-electric-vehicles.html

29. Goodall, W., Dovey, T., Bornstein, J., and Bonthron, B., "The Rise of Mobility as a Service," *Deloitte Rev* 20, no. 1 (2017): 112-129.

30. Airmee Website, "Discover a New Standard of Last-Mile Delivery," accessed February 2022, https://www.airmee.com

31. Youtube, "MUJI無印良品:GACHA Autonomous Shuttle Bus," accessed August 2022, https://youtu.be/W0oxz9KpiIY

SECTION 4

Disruptions, Challenges, and Benefits of ACES

Disruptions Caused by ACES Mobility

7.1. Background

What does it imply for a technology to be disruptive? This concern is briefly discussed in one of the author's previous works. The paper titled "Emerging Technologies with Disruptive Effects: A Review" documented the emerging technologies relative to the FIR and concisely noted the impact and disruptions that they will cause [1]. Innovation itself is divided into two types, i.e., sustaining and disruptive. Sustaining innovation refers to the creation which delivers the innovative elements via existing product portfolios and ecosystems. Disruptive innovation, on the other hand, is an innovation that produces new technological and market ecosystems, thereby replacing the incumbent technologies [2]. As has been defined and established in the previous chapters, ACES mobility is a disruptive innovation that unsettles not only the automotive and mobility sectors by introducing pristine innovations but will also bear social, legal, and technological disruptions. This chapter discusses the disruptions caused by ACES mobility.

7.2. Social Disruptions

Social disruption refers to terminology from a social science perspective that describes the alteration of social aspects of a community. One of the factors for the mentioned event can be caused by technological innovations. The current and potential social disruptions that will be brought about by ACES can be found in the following subsections.

7.2.1. Work-Life Balance Improvements

In major cities, traffic jams are not uncommon. It is not unusual for a city dweller to go out from the home to the workplace as early as two hours before the start of working hours. The same hours are also applicable for return trips, which means 4 hours of commuting time daily in some cases. Usually, this is caused by traffic jams. One of the potential benefits of ACES technology (particularly via the means of vehicle automation) is the reduction of traffic jams [3]. With good business and technical strategies to deliver ACES as a new mobility solution through robotaxis, the AD shuttles can be deployed in downtown areas, which means decreased vehicle usage and a reduction in the total number of vehicles downtown. If implemented, traffic jams downtown can be reduced, thereby improving the quality of work-life balance for urban residents in big cities by not having to be in traffic jams (Figure 7.1).

7.2.2. Disruptions for Media and Infotainment

If you are a frequent traveler to different countries, you will notice that in-flight entertainment for aircraft passengers has improved significantly in recent years. For example, in-flight entertainment now includes not only music and movies but also games and live news updates. The same growth will be seen because of the disruptions caused by ACES. With ACES as the future mobility, the vehicle will be seen not only as a platform but also as a "travel space," similar to aircraft. Therefore, it is expected that for driverless vehicles, the in-vehicle infotainment will also be transformed with more content such as movies and music, which is also made possible by the vehicle connectivity elements [4, 5] (Figure 7.2).

FIGURE 7.1 Shared driverless mobility of ACES has the potential to reduce traffic jams, subsequently improving the work-life balance of city dwellers.

Sasin Paraksa/Shutterstock.com.

FIGURE 7.2 What kind of infotainment will be there with regard to ACES mobility?

Andrey Suslov/Shutterstock.com.

7.2.3. Merging with the Other Industrial Revolution 4.0 Ongoing Developments

The discussions of ACES are not standalone. In fact, it is heavily interconnected with other ongoing innovations in different various sectors. The driverless vehicle technology will, for example, help to push forward a couple of other disruptive technologies such as blockchain and autonomous VTOL technology by either integration into its technology modules or adoption of a similar business strategy [6, 7]. 5G productization will also be expedited by incorporating the technology with vehicular connectivity modules of ACES mobility [8]. Apart from that, the electrification and battery industry will also be stimulated by the EV sector of the automotive industry [9]. It is apparent from this paragraph that ACES technology will complement the growth of the FIR.

7.2.4. Redefining Mobility

Some might argue, *"Do we even need to use cars for short trips?"* *"Yes, we have kids,"* might be the answer from others. *"But do we really need to own separate vehicles?"* asks another person. *"Can't we just share them?"*

Good questions indeed. We have been accustomed and subconsciously conditioned to the definition of mobility that has been molded over the past decades. With the growth of sharing economy, user behavior is also transforming [10]. Companies like Uber, Lynk & Co., and Lyft are pioneering this disruption with distinctive business models, allowing the redefinition of mobility. Consequently, in cities globally, today one does not necessarily need to own a vehicle, rather one can order vehicles from on-demand mobility mobile apps [11]. With ACES, the disruptions will be more substantial as it will also see a better quality of the shared mobility vehicle platform.

7.2.5. Disrupting Future Urban Planning

Major global cities have been congested and one of the main factors of the crowdedness is the "ad hoc" road network constructions.

With proper planning of ACES, instead of improving the roads in the cities reactionarily to the traffic jam demands, the technology has the potential to influence better future urban planning, where shared mobility will be the core foundation. For example, with the reduction of private vehicle ownership, cycling lanes can be increased instead of parking lots. In fact, countries like the Netherlands have successfully transformed their urban infrastructure mainly by the adoption of bicycles and reducing inner-city vehicles [12]. With ACES, this can be further increased and improved (Figure 7.3).

Furthermore, with the expected decline of private vehicle ownership, ACES will foresee the reduction of highway constructions, therefore saving city areas that can be used for other applications such as agriculture [13]. New residential units can also be built with a fewer number of parking lots required, thereby potentially democratizing the costs of private property ownership. As can be seen, ACES will have the potential to influence better urban planning in the future with the redefined mobility. Similar to the previously disruptive technology of social media, ACES will also bring disruption to the social lifestyle.

FIGURE 7.3 ACES as future mobility have the potential to influence future urban planning.

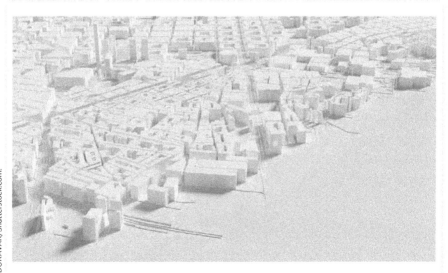

ADOKAVAK/Shutterstock.com.

7.3. Legal, Economical, and Workforce Disruptions

ACES, as a whole ecosystem, will bring disruptions to the legal and economic perspectives as well as the labor force. For example, the discussions on insurance, taxation, and regulations for ACES are among the foreseeable disruptions. Details are as follows.

7.3.1. Insurance Industry

ACES will see the removal of humans in the driving equation of AVs. This means the disruption will be on determining who is accountable should there be accidents or road fatalities. For example, if an AV is involved in an accident with other vehicles, who should be held responsible? Furthermore, the insurance industry must also participate in defining the operational and service-related activities for ACES. This means that ACES will be disrupting the said industry. In fact, several companies like Koop Technologies have been working to redefine the autonomy for automotive and mobility insurance [14]. However, this insurance disruption requires close collaboration between stakeholders in different sectors, which leads to another disruption from an ACES-related business perspective.

7.3.2. Taxations

Many countries around the world levy taxes on the use of vehicles on public roads. Road vehicles here include cars and motorcycles, among many others. However, as mentioned, ACES will see a reduction in vehicle ownership. Therefore, the new type of taxation must also be specified to compensate for the potentially reduced number of private vehicles. This includes the question of who should be taxed, which will then transform the automotive and mobility taxations. With the right approach, taxation not only can be innovated but also strengthened, where taxation can also include carmakers, end users, and other relevant entities in the ACES

FIGURE 7.4   New taxations will be seen with ACES mobility.

Funtap/Shutterstock.com.

ecosystems. The tax can then be utilized to expedite the needed infrastructure to facilitate the wider deployment of ACES. More discussions on taxations for ACES can be found in works by Ratner [15], Kovacev [16], and Fox [17] (Figure 7.4).

7.3.3. New Required Skills for Workforce

The new technologies brought by ACES innovations necessitate an army of new workforces with various skillsets. For example, compared to the traditional automotive industry, AVs witnessed the greatest demands in terms of the need for background experts in software programming. Furthermore, as the complexity of the field increases, the needs are not only limited to a computer science (CS) background, but recently, the demand also sees job listings mandating experts with CS background in addition to being proficient in back-end algorithms. Therefore, this creates a race of talent hiring among companies in the ACES sector. Apart from these skills, the disruption is also reflected in the need for multidisciplinary business experts to further develop the industry, including experts from digitalization topics such as Xaas [18].

7.3.4. New Job Ecosystems Opportunities

Not only do future employees need to prepare themselves with the new required skills, but they will also be presented with a wider choice of careers. For example, the control engineering-background experts can work not only as "vehicular platform control systems" engineers but can also now have opportunities to work on the "higher-level control systems" of AVs with topics such as "trajectory tracking," "motion planning," and "behavior planning." Therefore, it is a win-win situation that will benefit those who are preparing themselves for the future. Apart from that, as the ACES involves the outputs of new technologies, new jobs will also be created accordingly. For example, as companies like Einride, Sensible 4, and Ottopia concentrate on "Remote Monitoring for Autonomous Vehicle," this will create new types of jobs in the market.

7.3.5. Who Will Have the Ownership of ACES Vehicles?

Shared mobility provides a glimpse of the future with the concept of carsharing. It testifies that cars will not only be owned by private proprietors, but instead, by viewing the vehicle as a mobility platform, companies like Uber and Grab have managed to provide MaaS to their customers. As the industry continues to mature, the topic of "who actually owns the robotaxis" is an example of one of the disruptions caused by ACES [19].

7.3.6. Indirect Influence on Popular Culture

Media has always been influenced by current affairs as well as science and technology. For example, a lot of the most popular franchises in the entertainment industry are inspired by science and technology. Even in recent years, a lot of comedy series have started to show a glimpse of the more accurate version of the "lives of programmers and the tech industry in Silicon Valley," compared to the earlier version of "scientists with white coats in laboratories." With ACES as the future mobility, it is expected that more disruptions will be noticed in the content in the entertainment industry, which will eventually help to open up more discussions among the public about the said technology.

7.3.7. Passenger Behavior during the ACES Journey

As ACES witnesses the removal of human driving from the loop, it allows opportunities to rethink the concept of passenger behavior during the ACES ride. For example, in many autonomous shuttle prototypes, the internal design of the passenger seating is assembled differently compared to conventional cars. The disruptions that may be seen in the vehicles can be likened to those experienced by aircraft passengers during flight. Passengers will be able to sleep more comfortably in the vehicle (with the distinctive seating design) and watch movies together, or even a small cafe can be set up in autonomous shuttles for longer journeys. Some of these concepts have been envisioned by the Japanese company MUJI when collaborating with the Finnish company Sensible 4 [20]. Some researchers have also started studying the potential multimodality issues related to passenger behavior during the AD experience [21, 22, 23].

7.3.8. Regulations... and More Regulations

It could be argued that one of the reasons still hampering the progress of ACES mobility, particularly the AD features, is the homologation concerns of the innovation. This situation then caused a "limbo-like" scenario where the industry is trying to bring the technology; however, because of the lack of regulations available, it is being put "on hold." However, as interest in ACES increases again, in recent years a lot of noteworthy regulations, standards, and recommendations, among many others, have been outlined relating to ACES. The examples are:

- **Automated Vehicles: Summary of Joint Report, Summary of LC Report No 404 / SLC Report No 258** by the UK Law Commission [24]

- The document examines the future laws for AVs, where one of the central themes states that "the blame" should not be placed solely on human operators or drivers, but because it is the software that is "driving," the corresponding company should also be held accountable. Other themes are safety and accessibility for older and disabled people.

- **SAE J3131 Definitions for Terms Related to Automated Driving Systems Reference Architecture by the SAE International** [25]

  - Details the back-end components in the automated driving systems, according to SAE J3016, Taxonomy and Definitions for Terms Related to Driving Automation Systems for On-Road Motor Vehicles.

- **Data Act by the European Union** [26]

  - As ACES anticipates a lot of data shared by users to the corresponding organizations, it is logical that a lot of data privacy and security-related regulations have been drafted globally.

As can be seen from the examples above, the regulations not only involve technical discussions but cover different perspectives such as legal, technical, and cybersecurity. Further disruptions to regulations are expected to be caused by ACES, and not just in the automotive and mobility industries.

7.3.9. Private-Public-People Partnerships

To bring the mass deployment of new emerging technologies, particularly one that is conditional and heavily dependent on infra-structures like ACES, a very strong collaboration among private entities (e.g., technology companies), public organizations (e.g., government bodies), and people (e.g., the public) is needed. Therefore, ACES will disrupt the association between these entities in the years ahead to facilitate the productization of the technology. This involves cooperation and collaboration between cities and the industry players, among others [27, 28] (Figure 7.5).

FIGURE 7.5 Collaborations among different stakeholders are expected to increase in the coming years to facilitate ACES productization.

Lightspring/Shutterstock.com.

7.3.10. Increased Requirements of Empathy from the Leaders to Employees

Delivering and productizing emerging technologies is not easy. A lot of surveys have been done to find out about the stress among programmers for new innovations [29, 30, 31]. Therefore, one of the most important things that will be disrupted in the automotive and mobility industry for this topic is the requirement of leadership with high empathy, which focuses on the people and culture first, while ensuring business profits. By doing this, talent retention will be assured. This is very important as the Great Resignation, which is ongoing since early 2020 (as of early 2022), saw topics such as "moral injury" and "burnout" among employees to be examined [32, 33].

7.4. Technical, Technological, and Industrial Disruptions

It is the intention of the author of this book to assign the technical and technological disruptions as the "last" section discussion for the "disruptions by ACES," preceded by "Social Disruptions" and "Legal, Economical, and Workforce Disruptions." This is to accentuate that the major disruptions are not actually on the technical and technological but more on the non-technical aspects. However, in this section, the technical disruptions that will be caused by the new innovations of ACES are highlighted as follows.

7.4.1. ACES Increases the Software Importance in the Automotive and Mobility Industry

This has been reiterated a lot of times in the book; however, as some of the readers might only be reading this chapter, it is worth mentioning. Similar to other novel innovation elements spurred by the FIR, ACES will be catalyzed by the increased amount of software in its technology. As such, the term "software-defined" vehicles will be widely used to picture the future in which the value of a vehicle will be determined based on its software quality and complexity compared to only its hardware platform [34, 35].

7.4.2. Changes in the Business Models

Automakers will no longer be able to sustain their current business model to remain relevant in the complex future of the ACES industry. As ACES is a highly disruptive emerging technology, modifications and adaptations need to be done not only on the organizational and technical level but also at the business level. ACES will disrupt the automotive business similar to what Spotify and Netflix, among many others, have done to the music and entertainment industry [36]. To avert the collapse of automakers, the top management of the giants in the said industry should learn from the missteps and blunders of "previously collapsed" giants in other industries, such as during the early stages of the introduction of digital cameras and smartphones [37].

7.4.3. Process and Regulations Changes

As the technology is driven by software, the work processes within the industry that employed engineers and experts will also be disrupted.

As such, we have seen a lot of discussions on "agile software development" (for example) in the automotive industry in recent years. This requires a lot of training to be done within the organization to enable the safe deployment of ACES technology. More on this topic has been discussed throughout other chapters in the book.

7.4.4. Infrastructural Disruptions

Some of the elements of AD require the upgrade of the infrastructure. For example, the upgrades of traffic lights in the cities with the connectivity capability will help in improving the ACES safety where features such as V2X can be utilized to improve the information sharing between the vehicle and infrastructure (Figure 7.6). More details can be found in Chapter 4.

FIGURE 7.6   Infrastructure changes will also be seen to facilitate the wide deployment of ACES.

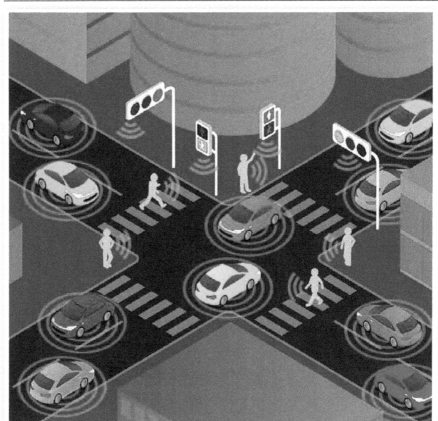

metamorworks/Shutterstock.com.

7.4.5. New Automotive and Mobility Industrial Stakeholders

ACES as the future mobility will also see the arrival of the new type of players in the said ecosystem. For example, for vehicle electrification, we are seeing a new supply chain of battery manufacturers in the automotive industry such as Northvolt, among many others. With the growing complexity of ACES, it is expected that more of such new industrial stakeholders and players will appear in the aforementioned ecosystems [38].

7.4.6. Democratizing Mobility

In most countries, as a standalone individual vehicle, ADAS technology like AEB has rarely benefited the lower-income segments of society as, usually, it comes with a higher price. The shared driverless mobility concept of ACES allows for the technology to be democratized, i.e., opening access to more society segments to benefit from the active safety innovations of ACES.

7.4.7. New Incomes for Countries

ACES not only will disrupt the business, but also the source of income for countries. For example, oil-supplying countries like Norway and United Arab Emirates (UAE) need to reexamine their reliance on oil to have a future-proof sustainable economy. Unsurprisingly, Norway and UAE are among those countries with the most serious efforts in supporting ACES. Efforts such as "Dubai Self-Driving Strategy" in Dubai and strategies outlined by The Norwegian Electric Vehicle Association in Norway are examples [39, 40, 41] (Figure 7.7).

7.4.8. Skunkworks Project-Organizations

As the nature of ACES is still uncertain, a lot of big carmakers are setting up spin-off companies with the aim to boost innovation within the said field. Skunkworks project is a term that refers to an establishment of a group within larger organizations to support rapid innovation [42]. Companies like Volkswagen have created

FIGURE 7.7 ACES via vehicle electrification sees the oil-producing countries start planning toward a more sustainable economy.

mentalmind/Shutterstock.com.

similar organizations like CARIAD to support the digitalization of their automotive business, with a focus on software. By doing so, automakers expect to achieve mandatory ambidexterity to remain relevant for future business [43].

7.4.9. More Concentrated Efforts to Tackle Climate Change Issues

ACES via vehicle electrification boost the discussions to reduce the climate change issues. With EVs, ACES not only provide the user with electrified mobility but also encouraged more discussions from academia on the topic of climate change, sustainability, and circular economies. Consequently, a lot of major carmakers are now moving in this direction, with some aiming to phase out ICE-based vehicles from their product lines soon [44].

7.5. Summary

In this chapter, the author lists the potential and ongoing disruptions that will be caused by ACES. The disruptions that were noted are from the "social aspects"; "legal, economical, and workforce disruptions"; and the "technical and industrial disruptions." It is apparent from the discussions in this chapter that ACES will bring a lot of disruptions to not only the automotive and mobility industry but also the whole society, if executed properly. In the next chapter, the challenges that will be faced by the different stakeholders in productizing ACES technology are discussed.

References

1. Rahman, A., Airini, U.Z.A.H., and Chin, T.A., "Emerging Technologies with Disruptive Effects: A Review," *Perintis eJournal* 7, no. 2 (2017): 111-128.

2. Christensen, C.M., *The Innovator's Dilemma: When New Technologies Cause Great Firms to Fail* (Boston, MA: Harvard Business Review Press, 2013)

3. UC Berkeley, "Eliminating Traffic Jams with Self-Driving Cars," accessed April 2022, https://ce.berkeley.edu/news/2537

4. Muguro, J.K., Laksono, P.W., Sasatake, Y., Matsushita, K. et al., "User Monitoring in Autonomous Driving System Using Gamified Task: A Case for VR/AR In-Car Gaming," *Multimodal Technologies and Interaction* 5, no. 8 (2021): 40.

5. Meurer, J., Pakusch, C., Stevens, G., Randall, D. et al., "A Wizard of Oz Study on Passengers' Experiences of a Robo-Taxi Service in Real-Life Settings," in *Proceedings of the 2020 ACM Designing Interactive Systems Conference*, Eindhoven, the Netherlands, 1365-1377, 2020.

6. Rathee, G., Sharma, A., Iqbal, R., Aloqaily, M. et al., "A Blockchain Framework for Securing Connected and Autonomous Vehicles," *Sensors* 19, no. 14 (2019): 3165.

7. Bloss, R., "By Air, Land and Sea, the Unmanned Vehicles Are Coming," *Industrial Robot: An International Journal* 34, no. 1 (2007): 12-16.

8. Tahir, M.N. and Katz, M., "ITS Performance Evaluation in Direct Short-Range Communication (IEEE 802.11 p) and Cellular Network (5G)(TCP vs UDP)," in Hamid, U.Z.A. and Al-Turjman, F. (eds.), *Towards Connected and Autonomous Vehicle Highways* (Cham: Springer, 2021), 257-279.

9. Liiv, O., "Industrialization of Lithium-Ion Prismatic Battery Cell for the Automotive Industry," 2020.

10. Standing, C., Standing, S., and Biermann, S., "The Implications of the Sharing Economy for Transport," *Transport Reviews* 39, no. 2 (2019): 226-242.

11. Casquero, D., Monzon, A., García, M., and Martinez, O., "Key Elements of Mobility Apps for Improving Urban Travel Patterns: A Literature Review," *Future Transportation* 2, no. 1 (2022): 1-23.

12. te Brömmelstroet, M., Boterman, W., and Kuipers, G., "How Culture Shapes-And Is Shaped By-Mobility: Cycling Transitions in the Netherlands," in Curtis, C. (ed.), *Handbook of Sustainable Transport* (Edward Elgar Publishing, 2020).

13. Davidson, P. and Spinoulas, A., "Autonomous Vehicles: What Could This Mean for the Future of Transport," in *Australian Institute of Traffic Planning and Management (AITPM) National Conference*, Brisbane, Queensland, 2015.

14. Koop Technologies Website, "Powering Autonomy Insurance," accessed February 2022, https://www.koop.ai

15. Ratner, S., "Taxation of Autonomous Vehicle in Cities and States," *Tax Law* 71 (2017): 1051.

16. Kovacev, R., "A Taxing Dilemma: Robot Taxes and the Challenges of Effective Taxation of AI, Automation and Robotics in the Fourth Industrial Revolution," *Ohio St. Tech. LJ* 16 (2020): 182.

17. Fox, W.F., "The Influence of Autonomous Vehicles on State Tax Revenues," *National Tax Journal* 73, no. 1 (2020): 199-234.

18. Coveri, A. and Zanfei, A., "Who Wins the Race for Knowledge-Based Competitiveness? Comparing European and North American FDI Patterns," *The Journal of Technology Transfer* (2022): 1-39, https://link.springer.com/article/10.1007/s10961-021-09911-z

19. Ward, J.W., Michalek, J.J., Samaras, C., Azevedo, I.L. et al., "The Impact of Uber and Lyft on Vehicle Ownership, Fuel Economy, and Transit across US Cities," *Iscience* 24, no. 1 (2021): 101933.

20. Youtube, "MUJI無印良品:GACHA Autonomous Shuttle Bus," accessed April 2022, https://youtu.be/W0oxz9KpiIY

21. Singleton, P.A., "Multimodal Travel-Based Multitasking during the Commute: Who Does What?" *International Journal of Sustainable Transportation* 14, no. 2 (2020): 150-162.

22. Hamadneh, J. and Esztergár-Kiss, D., "Multitasking Onboard of Conventional Transport Modes and Shared Autonomous Vehicles," *Transportation Research Interdisciplinary Perspectives* 12 (2021): 100505.

23. Hamadneh, J. and Esztergár-Kiss, D., "Evaluating Onboard Multitasking on Mode Choice: Conventional Car and Autonomous Vehicle," in *2021 7th International Conference on Models and Technologies for Intelligent Transportation Systems (MT-ITS)*, Heraklion, Greece, 1-6, IEEE, June 2021.

24. Scottish Law Comission, "Automated Vehicles: Joint Report," accessed April 2022, https://s3-eu-west-2.amazonaws.com/lawcom-prod-storage-11jsxou24uy7q/uploads/2022/01/Automated-vehicles-joint-report-cvr-24-01-22.pdf

25. SAE International, " Definitions for Terms Related to Automated Driving Systems Reference Architecture," accessed April 2022, https://www.sae.org/standards/content/j3131_202203/

26. European Commission, "Data Act: Proposal for a Regulation on Harmonised Rules on Fair Access to and Use of Data," accessed August 2022, https://digital-strategy.ec.europa.eu/en/library/data-act-proposal-regulation-harmonised-rules-fair-access-and-use-data

27. Majamaa, W., *The 4th P-People-In Urban Development Based on Public-Private-People Partnership* (Helsinki: Teknillinen korkeakoulu, 2008)

28. Pinkse, J., Bohnsack, R., and Kolk, A., "The Role of Public and Private Protection in Disruptive Innovation: The Automotive Industry and the Emergence of Low-Emission Vehicles," *Journal of Product Innovation Management* 31, no. 1 (2014): 43-60.

29. Kock, N., Moqbel, M., Jung, Y., and Syn, T., "Do Older Programmers Perform as Well as Young Ones? Exploring the Intermediate Effects of Stress and Programming Experience," *Cognition, Technology & Work* 20, no. 3 (2018): 489-504.

30. Milenović, M., Veljković, M., and Pešić, N., "Professional Styles and Some Psychophysiological Risk Factors of Work Engagement of Programmers," in Živković, S., Krstić, B., and Rađenović, T. (eds.), *Handbook of Research on Key Dimensions of Occupational Safety and Health Protection Management* (Hershey: IGI Global, 2022), 163-179.

31. Claes, M., Mäntylä, M.V., Kuutila, M., and Adams, B., "Do Programmers Work at Night or during the Weekend?," in *Proceedings of the 40th International Conference on Software Engineering*, Gothenburg, Sweden, 705-715, 2018.

32. Leslie, J.B., "Pandemic Paradoxes and How They Affect Your Workers," 2020, accessed April 2022, https://cclinnovation.org/wp-content/uploads/2021/10/pandemic-paradoxes-and-how-they-affect-your-workers.pdf

33. Sheather, J. and Slattery, D., "The Great Resignation—How Do We Support and Retain Staff Already Stretched to Their Limit?" *BMJ* 375 (2021): n2533.

34. Han, S., Cao, D., Li, L., Li, L. et al., "From Software-Defined Vehicles to Self-Driving Vehicles: A Report on CPSS-Based Parallel Driving," *IEEE Intelligent Transportation Systems Magazine* 11, no. 1 (2018): 6-14.

35. Zhao, F., Song, H., and Liu, Z., "Identification and Analysis of Key Technical Elements and Prospects for Software-Defined Vehicles," SAE Technical Paper 2022-01-7002, 2022, https://doi.org/10.4271/2022-01-7002

36. Nałęcz, A., "Cutting-Edge Business Models in the Age of Digital Disruption-Examples, Prospects, and Key Economic and Legal Challenges," PhD dissertation, University of Warsaw, 2019.

37. Metzler, F., "Firms, Industries, and Technological Change: A Patent-Based Approach to Studying Disruption and Disruptors," PhD dissertation, Massachusetts Institute of Technology, 2019.

38. Brenner, W. and Herrmann, A., "An Overview of Technology, Benefits and Impact of Automated and Autonomous Driving on the Automotive Industry," in Linnhoff-Popien, C., Schneider, R., and Zaddach, M. (eds.), *Digital Marketplaces Unleashed* (Berlin: Springer, 2018), 427-442.

39. Quartz, "Dubai Wants to Make 25% of Car Trips Driverless by 2030, and It's Starting with Tesla Taxis," accessed April 2022, https://qz.com/911900/dubai-wants-to-make-25-of-car-trips-driverless-by-2030-and-its-starting-with-tesla-taxis

40. Hafiz, D. and Zohdy, I., "The City Adaptation to the Autonomous Vehicles Implementation: Reimagining the Dubai City of Tomorrow," in Hamid, U.Z.A. and Al-Turjman, F. (eds.), *Towards Connected and Autonomous Vehicle Highways* (Cham: Springer, 2021), 27-41.

41. Norsk elbilforening, "The Norwegian Electric Vehicle Association," accessed April 2022, https://elbil.no/english/about-norwegian-ev-association

42. Oltra, V., Donada, C., and Alegre, J., "Facilitating Radical Innovation through Secret Technology-Oriented Skunkworks Projects: Implications for Human Resource Practices," *Human Resource Management Journal* 32, no. 1 (2022): 133-150.

43. Donada, C., Mothe, C., and Alegre, J., "Managing Skunkworks to Achieve Ambidexterity: The Robinson Crusoe Effect," *European Management Journal* 39, no. 2 (2021): 214-225.

44. Tsakalidis, A. and Thiel, C., "Electric Vehicles in Europe from 2010 to 2017: Is Full-Scale Commercialisation Beginning. An Overview of the Evolution of Electric Vehicles in Europe," EUR 29401, 2018.

8

Potential Challenges of ACES

8.1. Background

Productizing ACES technology is not as straightforward as it sounds. Delivering safe, reliable, and secure shared driverless mobility to the mass audience is not as simple as a "start-up pitch" makes it sound. In fact, the challenges that exist for this topic are enormous! Not only related organizations need to deliver new technology, enter the market, and assure reliability, but they also need to transform their own organizations while doing so. Hence, because of these challenges, no wonder the industry is seeing the entrance, exit, rise, and fall of the technology companies working on this topic at a rapid pace [1]. Despite that, the potential benefits that will be brought by ACES to humankind are huge, causing the investors to keep funding the ACES industrialization effort. In this chapter, the challenges of ACES development and productization are written, with the main targeted group of readers being the general audience. It follows the technical discussions in the earlier chapters. As such, this chapter will be written in a brief and concise manner to allow better understanding by the non-technical background readers. The author hopes that this chapter will also help draw a bigger picture of the ACES challenges to the researchers, therefore encouraging more works to improve and solve the problems.

8.2. Technical Challenges of ACES

8.2.1. Dealing with Uncertainties

Of course, the most prominent challenge for the ACES, particularly from the topic of AD is handling the uncertainties. The uncertainties in AD are intricate (Figure 8.1). Because of the multi-layer of the AD software, uncertainty can occur in the high-level modules such as perception, positioning, motion planning, and high-level control systems. The driverless platform should also be able to address the uncertainty in the low-level software which actuates the vehicle platform itself. Furthermore, in addition to that, each module should be able to operate reliably in various operational design domains in a very dynamic road traffic environment. Therefore, to assure the productization of the AD system is a feasible process, the scope of the development should be managed and constrained. This will then prevent scope creep

FIGURE 8.1   AD systems of ACES mobility needs to be able to handle the uncertainties from various aspects [3, 4].

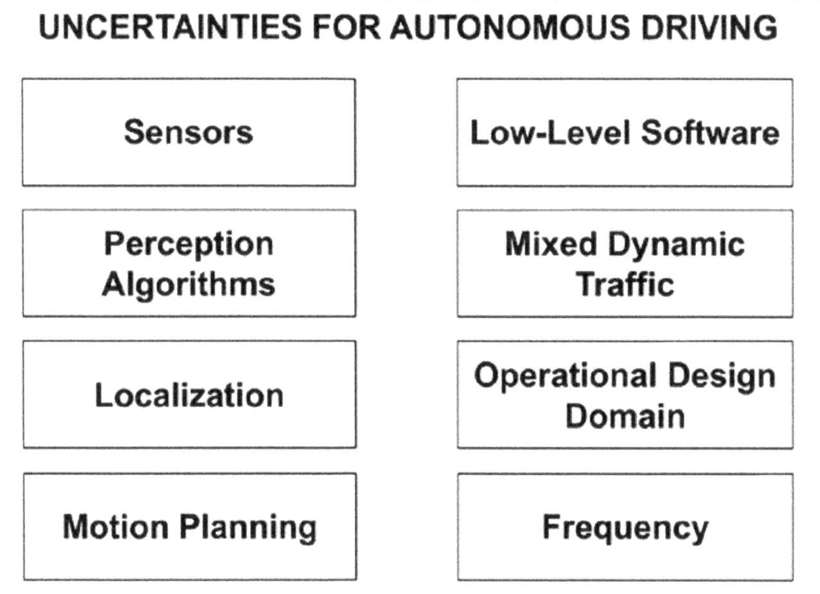

UNCERTAINTIES FOR AUTONOMOUS DRIVING

Sensors	Low-Level Software
Perception Algorithms	Mixed Dynamic Traffic
Localization	Operational Design Domain
Motion Planning	Frequency

© SAE International.

during the operational stage of the development, thereby preventing technical debt in the ACES software [2].

8.2.2. Network Latency

At the very core of CVs, the ability to sustain the connectivity elements throughout the whole journey is critical. However, how reliable is the connectivity performance under specific conditions and situations where there are network latency issues? This is one of the main challenges that should be assessed. Network latency concerns will cause delays not only in monitoring the ACES vehicle remotely but also in data transfer predicaments during potential hazardous scenarios. Fortunately, a lot of work has been done to address these concerns, which saw the progress made not only in 5G Internet but also in 4G and 6G for ACES applications [5, 6].

8.2.3. Cybersecurity

What will happen if the biggest social media website is taken down by hackers? What will happen to the data of the stated website users? What will happen to their privacy? For ACES, the "worst-case scenario" in the case of hacking or hijacking will be far worse compared to social media hacking. For example, a nonsecure ACES system might allow hackers to overtake vehicle control during an AD journey. This can cause fatalities if not prevented well. However, worry not as we are seeing more regulations-related discussions being made to address this topic.

Among notable outputs are "Recommendations for the Security of Connected and Automated Mobility" by ENISA, the Data Act by the European Union, and the California Consumer Privacy Act in the USA [7–9]. Although the intricacy of this topic will increase as the sophistication of the ACES industry grows, the recent updates show the growing awareness on this topic by the regulatory bodies.

8.2.4. Range Anxiety Is Still an Issue

Despite the efforts to increase the driving range of EVs in recent product lines by carmakers, range anxiety among end users is still visible. This is still apparent in developing countries, such as Malaysia, Thailand, and Indonesia [10, 11], as the lack of charging infrastructures remains an issue in promoting electrified mobility among the mass audience. Therefore, collaborations are needed between different stakeholders to improve the EV charging networks in developing countries.

8.2.5. Infrastructural Improvements to Support Vehicle Electrification

The reason why the adoption of ICE vehicles remains prevalent compared to EVs lies in the superior existence of gas station networks compared to the networks of EV charging facilities. In recent years, there have been a lot of discussions to improve the ecosystem of the electrified mobility charging infrastructures. Volvo Cars, for example, announced their effort to integrate their product lines with wireless charging technology [12]. Meanwhile, in China, a report by Bloomberg claims that battery swapping service for EVs is big in the said country [13]. Automotive and mobility industries are also witnessing more battery plants being built across the world to support EV industrialization, which eventually will prompt the growth of the ACES mobility sector [14].

8.2.6. Battery Recycling and Waste Management

As electrification and digitalization see the high dependence on battery-related technologies, it brings the question of "where do these batteries go after the end-of-life stage of their lifecycle?" Similar topics have been addressed with smartphone batteries [15]. For EVs of ACES mobility, this topic is of crucial importance. Comprehensive end-to-end battery lifecycle management is required to prevent the environmental impact of battery wastage [16] (Figure 8.2).

8.2.7. Hygiene Topics

COVID-19 has seen the rise of consciousness to "germs" and "hygiene" concerns among the mass audience. The sale of hand sanitizers increased since the initial phase of the said pandemic

FIGURE 8.2 Battery management and recycling topics should be considered to prevent environmental-effect from battery wastage.

asharkyu/Shutterstock.com.

globally [17]. As ACES encompass the shared mobility concept in its technology, the sanitation management of ACES technology is another challenge that needs to be considered by automotive ecosystems.

8.2.8. Ridesharing and Traffic Jams

Ironic as it seems, although originally introduced with the "notion" to reduce traffic jams by reducing "private vehicle ownerships," in some cities, ridesharing is shown to increase traffic jams [18]. Therefore, for ACES mobility to really benefit society, not only the technological developments need to be accomplished, but the regulations and systems should also be revised to support the objectives of future mobility. Therefore, instead of having private AVs for different passengers, what is potentially needed is an on-demand shared driverless mobility concept. This will then actually help to reduce the number of vehicles "on the road." However, this mandates collaboration between private and public stakeholders as it involves regulatory-related updates [19].

8.2.9. Traveling Salesman Problem

The traveling salesman problem refers to the question of determining the best optimal route for the desired journey. For ACES, as it will not be controlled by human drivers, the ACES should have the ability to formulate the best routes for every potential destination. This requires the integration of a good AD fleet management system and good mapping technology, among many others. It should also take the consideration of the live traffic jam information to allow maneuver rerouting by ACES vehicles [20, 21].

8.3. Legal, Industrial, and Workforce Challenges of ACES

8.3.1. Vandalism and Petty Crimes

A few years ago, the world witnessed the rise of the shared mobility sector. The growth did not happen only with companies like Uber, but it also occurred in other types of transportation such as bike sharing. However, in developing countries, because of the lack of regulations and awareness, a lot of vandalism has occurred with shared bicycles [22, 23]. To prevent this from happening for ACES, appropriate regulation should be made where perpetrators of vandalism could be penalized.

8.3.2. Whose Fault Is It? Who Is to be Blamed?

Removing the human element from the AD equation brings a high number of uncertainties into the ACES process. This opens up a myriad of questions particularly in the edge case scenarios during driverless navigation. For example, who will be blamed in the case of AV accidents? Is it the engineer or the manager? Is it the vehicle hardware manufacturer or software supplier? How to make sure that everything is traceable in the software pipeline—from the high-level AD software stack to the low-level vehicular platform? These are among the examples of how the complexity of ACES software is not something to be belittled. In fact, clear processes and regulations should be integrated into the development cycle to put the necessary responsibilities to everyone involved in delivering the product. Furthermore, this also raises concerns about insurance for ACES mobility.

8.3.3. Scope Creep and Technical Debt in the Software Development

The author has co-written an extensive paper about this topic, titled "Challenges of Complex Software Development for Emerging Technologies in Automotive Industry: Bridging the Gap of Knowledge Between the Industry Practitioners," published in the SAE World Congress 2022 Proceedings. The challenges with ACES software development are because, in certain parts of the industry, the development is purely driven by the *hype* where occasionally misleading info is being marketed. Consequently, this poses the risk of accidents because of scope creep that happens during development, which eventually causes technical debt in the software architecture. More details can be found in "Challenges of Complex Software Development for Emerging Technologies in Automotive Industry: Bridging the Gap of Knowledge Between the Industry Practitioners" [2].

8.3.4. Pricing Is Still Expensive

For a new technology to be accepted by the mass audience, it needs to be marketed and sold at an acceptable price range. However, as the productization of new technology is not cheap, usually emerging technologies will be initially offered at quite an expensive price. Consequently, this has slowed the acceptance of EV technology as the price is still not yet affordable (Figure 8.3). However, as we are seeing parallel growth in the ACES-related supply chains, the

FIGURE 8.3   Pricing remains one of the challenges to public acceptance of ACES.

impact of "economies of scale" on improving the efficiency of the productization of ACES technology will eventually help facilitate the affordable pricing of new innovations [24]. Furthermore, incentives should be given by the government to reduce other costs that need to be paid by the public, such as ticketing fees for shared driverless mobility. This will then prompt the user's acceptance of the said technology.

8.3.5. Unclear Requirements because of the Knowledge Gap in the Business-Facing Organizations

As mentioned in the previous point, ACES development is sometimes driven by *hype*. While it is good for attracting investments, however, this indirectly raises the number of customer-facing employees who do not have any prior knowledge about the complexity of ACES technologies. This causes selling something that is infeasible to be developed, which puts unnecessary stress on the developers. Thereby, in the mentioned paper by Tsakalidis and Thiel [24], the gap between business and technical experts on ACES should be bridged to prevent irresponsible ACES product development. Furthermore, this also necessitates a new type of leadership in the automotive sector to facilitate the transformation of the industry.

8.4. Social and Ethical Challenges of ACES

8.4.1. Changing Job Landscapes

A hundred years ago, the job landscape was different from what it is today. Partially, the changes are because of the distinct technologies which have been adopted by the mass public in their daily lives. For example, job expertise like software engineering were not highly in demand in the 1950s. This has changed with the arrival of new technological devices such as smartphones and laptops. One of the most critical challenges for ACES as the future mobility is that it will cause rapid demand for the necessary skills to facilitate the development. These include new automotive processes, technical skills, and business prowess. These will see the talent hiring races between companies to support their ACES ambitions. Apart from that, ACES will also demand that governments and public institutions around the world design the education systems accordingly to train the younger generation with the related skills. Fostering educational institutions to support lifelong learning will help accomplish this [25].

8.4.2. Transparency Is Needed

As ACES is disruptive in nature, it requires the talks of new standardizations and regulations by public and private institutions. This is needed to assure that ACES as mobility could be deployed to different countries with various challenges around the world. However, for regulations and standardizations to be proposed, transparency and a low level of corruption are required. This is because inefficient regulation development will not only cause accidents with ACES mobility (particularly in the case of AD) but also might cause fatalities [26, 27].

8.4.3. Requirement for a Visionary and Skillful Public Leadership

To bring ACES to the public, the competence to do so is not only necessary for technical organizations but also for public institutions. This demands the implementation of change management activities in public and government agencies to facilitate the deployment of new technologies in the automotive and mobility sectors. Therefore, young talents should be integrated into the strategic planning of governments, whether through partnership or employment. This requires the practice of nepotism to be at least reduced, while merit-based attitudes should be encouraged [28].

8.4.4. Importance of Education and Propagations

Vandalism as a potential challenge has been mentioned in one of the previous subsections. However, as it is related to technical, legal, and ethical concerns, collaborations among all related stakeholders are needed to prevent the vandalism of ACES vehicles by unsolicited perpetrators. For example, in addition to in-vehicle monitoring in ACES vehicles, regulations should be designed to penalize those who are involved in petty crimes such as vandalism on the said vehicles [29]. However, these are potentially still not enough. The members of the public deserve to receive enough information and education on ACES technology. Therefore, it is a necessity for government and private agencies to collaborate to educate the masses about the said technology and thereby curb the potential for vandalism (Figure 8.4). Continuous education to all segments of the members of society is important to prevent the gap in and disparity to access the ACES technology once it will be deployed.

8.4.5. False and Misleading Marketing

It is well known that ACES growth is partially spurred by the start-up sector. However, as start-ups require investment, misleading marketing is sometimes used, especially when it comes to emerging technology [30]. This can be deceptive as the

FIGURE 8.4 Stakeholders from different organizations should collaborate to prevent the potential vandalism of ACES vehicle platforms.

Imascaretti/Shutterstock.com.

technology might not be in the final stage of production yet, i.e., not yet reliable to be used. If poorly managed, this will create unnecessary pressure to deploy the product to the public, potentially causing fatalities because of technical errors. Apart from that, other terms such as "greenwashing" should also be monitored to prevent public misunderstanding of the ACES technology [31].

8.5. Summary

This chapter emphasizes potential challenges that will be brought about by productization of ACES technology. The author intentionally listed non-technical challenges for most of the sections to show that the ACES development will be facing not only technical issues but most important ethical, legal, and social challenges, among many others. Of course, other challenges that should be mentioned here are, for example, the abuse of the technologies, vaporware software marketing, and technical challenges such as sensor fusion and the back-end algorithms, among many others. However, the author believes the readers will get a glimpse of this topic from this chapter. In brief, it will take a lot of collaboration among different stakeholders to make the ACES technology widely available for public use.

References

1. Hawkins, A.J., "The Autonomous Vehicle World Is Shrinking—It's Overdue," The Verge, accessed April 2022, https://www.theverge.com/22423489/autonomous-vehicle-consolidation-acquisition-lyft-uber

2. Hamid, U.Z.A., Irimescu, D.S., and Zaman, M.T., "Challenges of Complex Software Development for Emerging Technologies in Automotive Industry: Bridging the Gap of Knowledge Between the Industry Practitioners," SAE Technical Paper 2022-01-0109, 2022, https://doi.org/10.4271/2022-01-0109

3. Carvalho, A., Gao, Y., Lefevre, S., and Borrelli, F., "Stochastic Predictive Control of Autonomous Vehicles in Uncertain Environments," in *12th International Symposium on Advanced Vehicle Control*, Seoul, South Korea, 712-719, 2014.

4. Koopman, P., Osyk, B., and Weast, J., "Autonomous Vehicles Meet the Physical World: RSS, Variability, Uncertainty, and Proving Safety," in Romanovsky, A., Troubitsyna, E., and Bitsch, F. (eds.), *International Conference on Computer Safety, Reliability, and Security* (Cham: Springer, 2019), 245-253.

5. Amoozadeh, M., Raghuramu, A., Chuah, C.N., Ghosal, D. et al., "Security Vulnerabilities of Connected Vehicle Streams and Their Impact on Cooperative Driving," *IEEE Communications Magazine* 53, no. 6 (2015): 126-132.

6. Kotilainen, I., Händel, C., Hamid, U.Z.A., Nykänen, L. et al., "Connected and Automated Driving in Snowy and Icy Conditions-Results of Four Field-Testing Activities Carried Out in Finland," *SAE Intl. J CAV* 4, no. 1 (2021): 109-118, doi:https://doi.org/10.4271/12-04-01-0009

7. European Union Agency for Cybersecurity, "Recommendations for the Security of CAM," accessed February 2022, https://www.enisa.europa.eu/publications/recommendations-for-the-security-of-cam

8. European Commission, "Data Act: Commission Proposes Measures for a Fair and Innovative Data Economy," accessed February 2022, https://ec.europa.eu/commission/presscorner/detail/en/ip_22_1113

9. Mesarčík, M., "Apply or Not to Apply? A Comparative View on Territorial Application of CCPA and GDPR," *Bratislava Law Review* 4, no. 2 (2020): 81-94.

10. Jamaludin, N.F., Hashim, H., Ho, W.S., Lim, L.K. et al., "Electric Vehicle Adoption in ASEAN; Prospect and Challenges," *Chemical Engineering Transactions* 89 (2021): 625-630.

11. Thananusak, T., Punnakitikashem, P., Tanthasith, S., and Kongarchapatara, B., "The Development of Electric Vehicle Charging Stations in Thailand: Policies, Players, and Key Issues (2015-2020)," *World Electric Vehicle Journal* 12, no. 1 (2020): 2.

12. Volvo Cars, "Volvo Cars Tests New Wireless Charging Technology," accessed March 2022, https://www.media.volvocars.com/global/en-gb/media/pressreleases/295720/volvo-cars-tests-new-wireless-charging-technology

13. Bloomberg, "Battery Swapping for EVs Is Big in China. Here's How It Works," accessed February 2022, https://www.bloomberg.com/news/articles/2022-01-24/battery-swapping-for-evs-is-big-in-china-here-s-how-it-works

14. Northvolt, "Volvo Cars and Northvolt to Open Gothenburg R&D Centre as Part of SEK 30bn Investment in Battery Development and Manufacturing," accessed February 2022, https://northvolt.com/articles/volvo-northvolt-dec2021

15. Zhao, Y., Pohl, O., Bhatt, A.I., Collis, G.E. et al., "A Review on Battery Market Trends, Second-Life Reuse, and Recycling," *Sustainable Chemistry* 2, no. 1 (2021): 167-205.

16. Harper, G., Sommerville, R., Kendrick, E., Driscoll, L. et al., "Recycling Lithium-Ion Batteries from Electric Vehicles," *Nature* 575, no. 7781 (2019): 75-86.

17. Prajapati, P., Desai, H., and Chandarana, C., "Hand Sanitizers as a Preventive Measure in COVID-19 Pandemic, Its Characteristics, and Harmful Effects: A Review," *Journal of the Egyptian Public Health Association* 97, no. 1 (2022): 1-9.

18. CNN, "Uber and Lyft Are Creating a Traffic Problem for Big Cities," accessed August 2022, https://money.cnn.com/2017/10/11/technology/future/ride-hailing-cities-public-transit/index.html

19. Dia, H. and Javanshour, F., "Autonomous Shared Mobility-on-Demand: Melbourne Pilot Simulation Study," *Transportation Research Procedia* 22 (2017): 285-296.

20. Palhares, R.A. and Araújo, M.C.B., "Vehicle Routing: Application of Travelling Salesman Problem in a Dairy," in *2018 IEEE International Conference on Industrial Engineering and Engineering Management (IEEM)*, Bangkok, Thailand, 1421-1425, IEEE, 2018.

21. Philippe, C., Adouane, L., Tsourdos, A., Shin, H.-S. et al., "Probability Collectives Algorithm Applied to Decentralized Intersection Coordination for Connected Autonomous Vehicles," in *2019 IEEE Intelligent Vehicles Symposium (IV)*, Paris, France, 1928-1934, IEEE, 2019.

22. Rosnan, H. and Abdullah, N.C., "An Exploratory Study of Bicycle Sharing in Malaysia," *Journal of ASIAN Behavioural Studies* 4, no. 12 (2019): 25-36.

23. Sun, Y., "Sharing and Riding: How the Dockless Bike Sharing Scheme in China Shapes the City," *Urban Science* 2, no. 3 (2018): 68.

24. Tsakalidis, A. and Thiel, C., "Electric Vehicles in Europe from 2010 to 2017: Is Full-Scale Commercialisation Beginning. An Overview of the Evolution of Electric Vehicles in Europe," EUR, 29401, 2018.

25. Youtube, "Lifelong Learning in Finland—An Expat's View | Umar Zakir Abdul Hamid | TEDxMetropoliaUniversity," accessed February 2022, https://youtu.be/IgIs9RZTgq4

26. Holcombe, R.G. and Boudreaux, C.J., "Regulation and Corruption," *Public Choice* 164, no. 1 (2015): 75-85.

27. Breen, M. and Gillanders, R., "Corruption, Institutions and Regulation," *Economics of Governance* 13, no. 3 (2012): 263-285.

28. Jones, R.G. and Stout, T., "Policing Nepotism and Cronyism without Losing the Value of Social Connection," *Industrial and Organizational Psychology* 8, no. 1 (2015): 2-12.

29. Biermann, H., Philipsen, R., Brell, T., and Ziefle, M., "Shut Up and Drive? User Requirements for Communication Services in Autonomous Driving," in Krömker, H. (ed.), *International Conference on Human-Computer Interaction* (Cham: Springer, 2020), 3-14.

30. Nuseir, M.T., "Impact of Misleading/False Advertisement to Consumer Behaviour," *International Journal of Economics and Business Research* 16, no. 4 (2018): 453-465.

31. Dixon, L., "Autonowashing: The Greenwashing of Vehicle Automation," *Transportation Research Interdisciplinary Perspectives* 5 (2020): 100113.

Potential Benefits of ACES

U p to this point in the book, readers are expected to have an understanding of the background to the development of ACES technology, which has been facilitated by the current FIR. Based on the discussions in Chapter 2 on the current trends pointing toward the ACES arrival, the transformation of the mobility sector is imminent. The discussions in Chapters 3–6 gave the book readers a glimpse into the ACES back-end. Yet the changes that will be and are driven by ACES are not without challenges, and therefore in Chapters 7 and 8, the author notes these challenges.

Based on Chapters 2 through 8, if ACES productization and development can be done according to the required homologation and safety considerations, it will bring a lot of benefits not only to the end users but potentially also to the automotive industry ecosystem as well as the society. In this chapter, the author lists the potential benefits brought about by ACES technology. To highlight the potential impact of ACES on human society as a whole, the discussions are made from broad perspectives, not just technical viewpoints.

9.1. Technological, Safety, and Security Benefits

ACES innovations will be able to bring benefits to people from different perspectives. In this section, the technological, safety, and security benefits from the progress and productization of ACES are identified.

9.1.1. Prompting Safer Automation in Other Industries

Hamid et al. [1] mentioned that one of the challenges in creating a secure AV software stack stems from the apparent knowledge gap between practitioners. Automotive, an industry that has been traditionally remote from complex software development, is seeing the import of CS-background programming experts. However, because of the rapid progress of the industry, the dissemination of knowledge about the necessary safety requirements can not yet be circulated. This brought challenges to AD productization in the past few years. However, if the considerations that are suggested in the previous chapters are addressed, and ACES developments are done according to the needed safety considerations, ACES not only will be able to output safer mobility but also help in promoting safe autonomous systems development across other applications.

In particular, ACES mobility maturation has expanded discussions on the standardization for safety-critical software development. For example, the "Guidelines for Mobility Data Sharing Governance and Contracting," issued by Mobility Data Collaborative and published by SAE Industry Technologies Consortia [2], highlights the guidelines on data sharing between different stakeholders of different organizations. This will help to encourage similar discussions in other types of autonomous systems development (Figure 9.1). ACES also promotes the importance of software best practices and process. For example, the growth of the field spurred discussions on the adoption of frameworks such as Automotive SPICE (Software Process Improvement and Capability dEtermination), also known as ASPICE, by automotive practitioners. This discussion will eventually reach other safety-critical autonomous systems such as robotics in healthcare, among many others [3].

9.1.2. Encouraging ACES Mobility across Different Transportation Domains

On most occasions, when discussing ACES mobility, the applications revolve around road vehicles. However, each of the elements

FIGURE 9.1 Safer automation across different applications can be inspired by the ACES-related safety discussions.

PopTika/Shutterstock.com.

in the ACES acronym has the potential to promote better mobility for other modes of transportation (Figure 9.2).

For example, as part of the shared future mobility, a successful AD business model will motivate other sectors such as the maritime industry to bring more autonomous surface vehicles and electric boats for public usage [4]. Eventually, mobility as a whole will not only be improved for road vehicles, but the transportation ecosystem will also have more variants for the modes of transportation.

Consequently, for countries with many islands geographically, ACES will not only help in solving the last-mile solutions for rural areas but also help bridge isolated islands to the mainland through maritime ACES implementation, which is inspired by the road vehicle use cases. In short, ACES progress for road vehicle applications, technologically and business-wise, will inspire more growth in ACES mobility for other transportation use cases.

9.1.3. Increased Safety

Of course, one of the main benefits of ACES will be the enhanced vehicle safety. In fact, one of the most popular motivations for vehicle automation is to reduce human error by the incorporation

FIGURE 9.2   ACES for road vehicles can stimulate the revolution in other transportation applications as well.

ilmarinfoto/Shutterstock.com.

of autonomous systems into the driving experience [5, 6]. However, as a caveat, the safety benefits of vehicular automation, particularly in reducing road fatalities, can only be realized if ACES development is done properly. If not, new kinds of accidents might occur because of software unreliability [7].

With ACES, vehicle automation will be able to provide AEB intervention on the appearance of a previously occluded object in the case of the carelessness of the human driver [8]. Reliable AD technology also creates a smoother highway cruising experience while maintaining safe distances from potential hazards, ensuring the safety of the passengers.

Vehicle connectivity will also help in improving vehicle safety with the V2X technology. With these capabilities, the robotaxi can not only avoid accidents with occluded physical objects, but it can also replan its trajectory in the occurrence of natural disasters such as flash floods with the V2X abilities. Furthermore, shared driverless mobility through the use of a robotaxi has the potential to offer parents peace of mind while their children are traveling to school, for example, by enabling the location-sharing features.

Despite this, just to recap, ACES will only be able to provide the safety benefit if it is designed and developed with the necessary homologation and process. More discussions on this can be found in the previous chapters.

9.1.4. More Transparency and Security for Mobility

Grab, a technology giant based in the Southeast Asian region (ASEAN), was formed initially to supply the market with an option of a safer ride home in lieu of taking the taxi service, which might be risky in a certain part of the city. Eventually, what was initially started as a taxi-booking app in Kuala Lumpur, Malaysia became the goliath of mobility in ASEAN. The essence here is to provide a more transparent record and data during the ridehailing-based journey [9].

With ACES, the connected and driverless shared electric transportation will provide more clarity for the mobility experience, therefore potentially increasing the safety of passengers during the ride home. For example, imagine the situation of a lady going home alone late at night after attending work events. With ACES mobility, she can be assured that the robotaxi is connected to the "main operations center" via "vehicle connectivity," and she can straightaway notify the remote operator in risky scenarios. Besides, with the user apps for booking the robotaxi, by including a regulation that mandates every user of the service to register, it is easy to trace the identity of any crime perpetrator (Figure 9.3).

Furthermore, with ACES, through a monitoring system that works similar to the aircraft black box, the recorded data for the AD driving can be used to monitor any criminal activity, therefore assuring end-user comfort during the navigation [10].

However, of course, this can only be achieved by addressing all the concerns regarding privacy and cybersecurity for CAM. Fortunately, a lot of regulations have been outlined for this topic, which includes the "Recommendations for the Security of CAM" by ENISA [11].

FIGURE 9.3 ACES can provide more security for mobility and transportation, especially during nighttime, due to their capability for monitoring and data recording.

Travel_Master/Shutterstock.com.

9.2. Societal and Sustainability Benefits

ACES not only brings benefits from a technical perspective. It also has the potential to bring benefits to society and the environment by providing a sustainable mobility ecosystem.

9.2.1. Lessened Car Ownership: A Chance to Reimagine Mobility

We have discussed the need for shared mobility in Chapters 2 through 8. Human society has become too accustomed (without we realize it or not) to having too many cars, sometimes for a significantly small population. Research shows that most cars are parked 95% of the time [12]. For example, on weekdays, most of us only use the car in the morning to get to the office and then again to drive home.

Therefore, with ACES, the shared mobility philosophy will allow for a reduction in private vehicle ownerships. Furthermore, there are also many developments in the industry that show that robotaxis are not designed according to the conventional vehicle layout but like autonomous shuttle vehicles [13, 14, 15]. With this shared philosophy and novel vehicle designs, humanity as a whole has a great opportunity to redefine the concept of mobility. Vehicle functionalities will have the opportunity to be reimagined for other purposes such as mobile tutoring centers and even mobile pizza restaurants, among many others. If this can be developed further, ACES mobility can also be used to provide aid to refugees in war zones or rural areas as mobile driverless clinics, for example.

9.2.2. Cleaner Mobility Ecosystem and Energy

With electricity as the backbone of ACES mobility, we will see lessened pollution in the cities. This will be very helpful in improving the quality of air, especially if ACES will be deployed in big metropolitans such as Istanbul, Tokyo, and Moscow (Figure 9.4).

Furthermore, the electrified future mobility of ACES will not only bring benefits for the lesser vehicle emission, but it will also provide smarter usage of spaces with EV charging spots. As can already be seen in major cities around the world, charging spots for EVs can be deployed in compact parking areas, whereas gas stations usually need a considerable amount of space. This will improve the city planning as well.

However, it is important to note that for EVs to successfully aid in achieving this benefit, the source of electricity should be clean too.

9.2.3. Improved Traffic in Cities

Traffic jams not only cause stress and headache for city dwellers worldwide, but instead, it also cost them huge fortunes, with a report by CNBC and the Texas A&M Transportation Institute citing the value of $179 billion each year for the costs of traffic jams in the USA [16].

FIGURE 9.4   EVs will help to provide a cleaner ecosystem for the mobility sector.

Kozak_studio/Shutterstock.com.

As mentioned previously, ACES has the enormous potential to reduce private vehicle ownership in large cities because of shared mobility which is a part of the disruptive technologies. With reduced vehicle ownership and an improved mobile experience through driverless transportation, the costs that need to be endured because of pointless traffic jams can be saved and utilized instead for other purposes of development. Indirectly, this will also boost the economy of nations through financial savings from the reduced traffic jams.

9.2.4. More Spaces for Urban Recreational Zones and Activities: Improving the City Attractiveness

Many reports have highlighted the impacts and benefits of reducing private vehicle ownership. Cities such as Paris, Helsinki, and

Amsterdam, for example, have been putting extra efforts into building bike lanes to encourage the reduction of private vehicle use in crowded areas. However, in some countries, cycling may not be the best solution for transportation because of the tropical rain and cold weather.

As previously mentioned, ACES has the potential to advance this effort by promoting shared driverless mobility. With this, the spaces that are now occupied as parking lots and highways, for example, can be adapted for more greenery and plant life in urban areas [17]. This will then allow for the reinvigoration of the cities and create potential tourism opportunities with the improved attractiveness and image of the city.

9.2.5. Less Stress for City Dwellers

If you have been living and working in big cities, for example, in some Asian countries, you will be able to vividly comprehend that "traffic jams cause stress." For example, hours of commuting yields unwarranted fatigue for employees even before reaching the office. This produces employees with low productivity for companies.

With ACES, shared mobility will complement the current public transportation, providing a last-mile solution with an enjoyable experience. With fewer private vehicles on the road, the traffic jam will be decreased, which can also help reduce the stress level among city residents, thereby improving the quality of life (Figure 9.5).

9.2.6. Potential to Reduce Crimes

One of the unexpected benefits of ACES mobility is its potential to reduce a certain crime level. The author of this book originally came from a developing country where "hit-and-run" is not unknown.

What is "hit-and-run"? For oversimplification, hit-and-run refers to an incident where the offender does not stop in the event of a collision. This has since sparked the interest in the use of "dashcams" by private users in their vehicles [18]. However, a

FIGURE 9.5   Shared mobility as part of the ACES ecosystem will potentially be able to reduce traffic jams in the big cities, therefore, reducing the stress levels of the city residents.

bibiphoto/Shutterstock.com.

dashcam is only helpful if the accident involves cars. In road accidents involving motorcyclists, where the perpetrator may be driving a car, a hit-and-run can still happen. In such cases, the victim of the accident may be left for dead until someone finds them because the perpetrator ran away after hitting the motorcycle (Figure 9.6).

With vehicle connectivity elements in ACES, these incidents can be reduced. By authorizing the V2V functionality, the incidents can be recorded, and the identity of the perpetrators can be uploaded and traced immediately from the authority's database. The author realized this potential benefit while writing this book and sincerely hopes for wider implementation of ACES and vehicle connectivity features (at the very least) for a safer road for everyone.

9.2.7. Better Social Security Benefits

Other indirect benefits from the reduction in private vehicle ownerships brought about by the implementation of ACES are the potential to reduce road maintenance. In big cities, because of the high

FIGURE 9.6 Hit-and-run crimes can potentially be reduced with broader adoption of vehicle connectivity.

yamasan0708/Shutterstock.com.

volume of traffic and vehicles on the road, efforts are needed to maintain the road by constantly resurfacing it. This resurfacing and maintenance costs a lot of money [19].

With ACES, private vehicle ownership can be reduced, therefore reducing the need for constant road resurfacing. The tax savings can then be used by the government to provide better social security, for example, by building bigger schools and daycares.

9.2.8. Facilitating Some Objectives of the UN SDGs

In the earlier chapters of this book, it is mentioned that the formulation of the UN SDGs has driven many innovations based on sustainable development. ACES as the future mobility, i.e., an autonomous, connected, electric, and shared mode of transportation, have the potential to facilitate some of the UN SDGs objectives. For example, the 11th element of the SDGs "sustainable cities and communities" can be achieved with ACES mobility by reducing vehicle ownership, therefore reducing vehicle emission.

FIGURE 9.7 If executed properly, ACES has the potential to help achieve the UN SDGs objectives.

yukipon/Shutterstock.com.

Furthermore, apart from addressing "climate change" (13th element of the SDGs) with vehicle electrification, ACES will also be able to indirectly help the 10th element of the SDGs, i.e., "reduced inequalities." As ideal as it may sound, ACES, if executed properly, particularly in the case of autonomous shuttle buses, has the potential to democratize mobility by providing everyone in cities and even rural areas equal access to mobility (Figure 9.7).

9.2.9. Improving Air Quality, Improving Health

According to a report by Anadolu Agency, a Turkish state-run news agency, there have been a lot of deaths in Bosnia and Herzegovina caused by problems related to air pollution [20]. According to the World Health Organization, air pollution can cause diseases such as lung cancer and acute respiratory infections.

ACES have the potential to reduce vehicle ownership with shared mobility. Furthermore, with electricity as the main catalyzer for vehicle energy, ACES, if deployed in air-polluted countries, will

FIGURE 9.8 ACES have the potential to improve air quality and reduce noise pollution in cities worldwide.

testing/Shutterstock.com.

be able to improve the air quality and indirectly the health quality, thereby improving the health of the residents in those areas, particularly older people as well as children. Furthermore, with the adoption of EVs by ACES, the noise in major cities worldwide can also be reduced (Figure 9.8).

9.3. Mobility User Benefits

ACES not only brings the aforementioned benefits; most importantly, it will also bring a new type of mobility for the public. In this section, the user benefits derived from the new mode of transportation are identified.

9.3.1. Better Transportations for the Society

The potentials of integrating ACES into the city transportation systems are huge (and the prospects are continuously growing). A project called LuxTurrim5G held in Finland has shown the potential of autonomous shuttle buses as a new method of public transportation. One of their public demonstration videos shows that the autonomous shuttle bus used in the project can be used for family recreational activity purposes [21].

ACES mobility applications can also be extended to provide a safer school bus journey. With the equipped ADAS in the AV platform, the robotaxi will be able to detect its surroundings and provide blind-spot monitoring, therefore reducing the potential of accidents between the bus and the school kids [22].

Apart from that, with ACES mobility a better reach of public transportation can be achieved, particularly in the rural areas where previously it might be hard to deploy drivers to the said area.

9.3.2. Reduced Costs of Transportation

Similar to other types of public transportation, ACES will not only be able to yield a cleaner environment but also enable end users to reduce their monthly costs of transportation. As ACES will be the complementary element of the current public transportation, this will also motivate society to use public transportation.

9.4. Economic Benefits

ACES will not only bring benefit to humankind, but it will also create a new automotive and mobility industry ecosystem. In this section, the author highlights some of the economic potential benefits.

9.4.1. New Investments Potential

The FIR brought with it an unlimited amount of opportunities. With new innovations created every year, it also brought a huge amount of investment for industrial growth.

Therefore, as we can see in the media today, big companies are racing to be the first to market full-stack ACES solutions. This will bring a lot of new job opportunities, especially to the younger generation.

Not only companies but nations of different countries are also racing to be the "testbed" and "center of the new mobility industry." In early 2022, Project Arrow was launched, which identifies the Canadian's ambition to lead the transformation from ICE to vehicle electrification [23].

ACES has also sparked the birth of many new start-ups by software engineering graduates. This development also brought in new processes and industrial mindsets. For example, contrary to the traditional industry, this new automotive industry encourages more transparency within the organization as well as co-creation for better innovations [24].

9.4.2. Opportunities for Developing Countries

ACES mobility will not only benefit developed nations economically. Since ACES depends on a lot of new sensors and external devices such as LiDAR, cameras, EV batteries, and microchips, it also unlocks opportunities for the manufacturing of these mentioned instruments.

Because of the lower costs of manufacturing in developing countries, we see many discussions on building manufacturing plants for ACES vehicle supplies in these countries. For example, several automotive companies have been planning to open EV battery plants in Southeast Asia [25].

9.4.3. New Business Models

ACES not only will disrupt the automotive industry but also disrupt the definition of business. For example, MaaS has managed to create a new type of economy where it also includes topics such as ridesharing as well as carpooling. Therefore, to be able to sustain profitability, carmakers should always reinvigorate their understanding of the technology.

Among the notable examples of new business models related to ACES are autonomous trucking by Einride as well as Gatik's collaboration with Walmart [26, 27].

9.4.4. Cross-Collaborations between Different Industries

ACES will also have the potential to disrupt the long-distance travel experience. With ACES, where connectivity and electricity are part of the mobility, better infotainment can be provided to the users during the trip. This will stimulate more collaborations between automotive companies with mobility providers as well as entertainment giants. This will then stimulate more economic benefits for the societies.

9.5. Summary

Finally, as can be seen from this chapter, ACES as the future mobility will bring benefits from different perspectives such as technological, societal, user experiences, and economic. The author intentionally emphasized the societal benefits to show that, by the end of the day, the disruptions that will be brought by ACES will be about "benefitting" us humankind.

References

1. Hamid, U.Z.A., Irimescu, D.S., and Zaman, M.T., "Challenges of Complex Software Development for Emerging Technologies in Automotive Industry: Bridging the Gap of Knowledge Between the Industry Practitioners," SAE Technical Paper 2022-01-0109, 2022, https://doi.org/10.4271/2022-01-0109

2. SAE International, " Guidelines for Mobility Data Sharing Governance and Contracting," accessed February 2022, https://www.sae.org/standards/content/mdc00001202004

3. Macher, G., Much, A., Riel, A., Messnarz, R. et al., "Automotive SPICE, Safety and Cybersecurity Integration," in Tonetta, S., Schoitsch, E., and Bitsch, F. (eds.), *International Conference on Computer Safety, Reliability, and Security* (Cham: Springer, 2017), 273-285.

4. Techwire Asia, "How Self-Driving Cars Are Inspiring Vessel Operations around Asia," accessed February 2022, https://techwireasia.com/2019/10/how-self-driving-cars-are-inspiring-vessel-operations-around-asia/

5. Sam, D., Velanganni, C., and Esther Evangelin, T., "A Vehicle Control System Using a Time Synchronized Hybrid VANET to Reduce Road Accidents Caused by Human Error," *Vehicular Communications* 6 (2016): 17-28.

6. Litman, T., *Autonomous Vehicle Implementation Predictions* (Victoria, Canada: Victoria Transport Policy Institute, 2017), 28.

7. Porathe, T., Hoem, Å., Rødseth, Ø., Fjørtoft, K. et al., "At Least as Safe as Manned Shipping? Autonomous Shipping, Safety and 'Human Error'," in Haugen, B.S., Barros, A., van Gulijk, C., Kongsvik, T. et al. (eds.), *Safety and Reliability-Safe Societies in a Changing World* (Boca Raton, FL: CRC Press, 2018), 417-425.

8. Hamid, U.Z.A., Zakuan, F.R.A., Zulkepli, K.A., Azmi, M.Z. et al., "Autonomous Emergency Braking System with Potential Field Risk Assessment for Frontal Collision Mitigation," in *2017 IEEE Conference on Systems, Process and Control (ICSPC)*, Melaka, Malaysia, 71-76, IEEE, 2017.

9. Bloomberg, " Harvard Inspires Man to Ditch Family Riches for Taxis," accessed February 2022, https://www.bloomberg.com/news/articles/2014-06-10/harvard-inspires-man-to-ditch-family-riches-for-taxis

10. Hamid, U.Z.A., Mehndiratta, M., and Adali, E., "Adopting Aviation Safety Knowledge into the Discussions of Safe Implementation of Connected and Autonomous Road Vehicles," SAE Technical Paper 2021-01-0074, 2021, https://doi.org/10.4271/2021-01-0074

11. European Union Agency for Cybersecurity, "Recommendations for the Security of CAM," accessed February 2022, https://www.enisa.europa.eu/publications/recommendations-for-the-security-of-cam

12. Morris, D.Z., "Today's Cars Are Parked 95% of the Time," *Fortune*, accessed February 2022, https://fortune.com/2016/03/13/cars-parked-95-percent-of-time

13. Sensible 4, "GACHA: The World's First Autonomous Shuttle Bus for All Weather Conditions, Launched in March 2019," accessed February 2022, https://sensible4.fi/gacha/

14. CNBC, "Cruise Expects GM to Begin Production of New Driverless Vehicle in Early 2023," accessed February 2022, https://www.cnbc.com/2021/05/13/gm-to-begin-production-of-new-driverless-cruise-vehicle-in-early-2023.html

15. Zoox Website, "The Future Is for Riders," accessed February 2022, https://www.zoox.com/

16. CNBC, "How Traffic Jams Cost the US Economy Billions of Dollars a Year," accessed February 2022, https://www.cnbc.com/2019/12/24/traffic-jams-how-they-form-and-end-up-costing-the-us-economy-billions.html

17. Queensland Government, "Benefits of Riding," accessed February 2022, https://www.qld.gov.au/transport/public/bicycle-riding/benefits-of-riding

18. Park, S., Kim, J., Mizouni, R., and Lee, U., "Motives and Concerns of Dashcam Video Sharing," in *Proceedings of the 2016 CHI Conference on Human Factors in Computing Systems*, San Jose, CA, 4758-4769, 2016.

19. Reger, D., Madanat, S., and Horvath, A., "Economically and Environmentally Informed Policy for Road Resurfacing: Tradeoffs between Costs and Greenhouse Gas Emissions," *Environmental Research Letters* 9, no. 10 (2014): 104020.

20. Anadolu Agency, "Bosnia Herzegovina Reports 20% of Deaths Caused by Air Pollution," accessed February 2022, https://www.aa.com.tr/en/environment/bosnia-herzegovina-reports-20-of-deaths-caused-by-air-pollution/2465210

21. Youtube, "Autonomous GACHA Bus Pilot in Nokia Campus," accessed January 2022, https://youtu.be/ZHYrejrnBww

22. Lemay, P. and Vincent, A., "School Bus Visibility: Driver's Field of View and Performance of Mirror Systems on a Conventional Long-Nosed School Bus," *SAE Transactions* 107 (1998): 1509-1529.

23. CNN, "Project Arrow: The All-Canadian, Zero-Emissions Electric Vehicle," accessed August 2022, https://sponsorcontent.cnn.com/int/invest-in-canada/project-arrow-the-all-canadian-zero-emissions-electric-vehicle/

24. Goncalves, D., Bergquist, M., Alänge, S., and Bunk, R., "How Co-Creation Supports Digital Innovation in Automotive Startups," in *29th European Conference on Information Systems: Human Values Crisis in a Digitizing World, ECIS 2021*, Marrakech, Morocco, June 14-16, 2021, Association for Information Systems, 2021.

25. Hamid, U.Z.A., Ishak, S.Z., and Imaduddin, F., "Current Landscape of the Automotive Field in the ASEAN Region: Case Study of Singapore, Malaysia and Indonesia—A Brief Overview," *Asean Journal of Automotive Technology* 1, no. 1 (2019): 21-28.

26. Sjoberg, K., "Automotive Industry Faces Challenges [Connected and Autonomous Vehicles]," *IEEE Vehicular Technology Magazine* 15, no. 3 (2020): 109-112.

27. Sjoberg, K., "Automated Trucks Overtake Self-Driving Cars [Connected and Automated Vehicles]," *IEEE Vehicular Technology Magazine* 17, no. 1 (2022): 94-97.

SECTION 5
Summary and Conclusions

10

Summary: "ACES Is Imminent. It Is a Bumpy Road. Cross-Organizational Collaborations Are a Necessity."

Some might be critical and perceive the vision of ACES as overrated. However, a lot of trends in R&D, industrial activities, and market opportunities indicate that it is happening, and will be an actuality. In fact, in the first two months of 2022 alone, news of at least five major investments related to the ACES topics with the potential amount of more than 1 billion USD are made in the said period [1, 2, 3, 4, 5]. Furthermore, big collaborations are continuously announced in the same period, such as Northvolt and Volvo's effort to build a gigafactory that will serve as a battery R&D center in Gothenburg [6].

Nevertheless, when one contemplates the skeptics' claim that ACES is overhyped, it is actually comprehensible for a lot of reasons. This book attempts to address and explain the "assumptions, speculations, and questions" surrounding the general audience on the theme "What exactly is Autonomous, Connected, Electric, and Shared Mobility from the big perspective?" and it is divided into five sections with a total of ten chapters.

The author has been a practicing expert in the field of ACES since 2014. With R&D and professional experiences in different countries and continents (Malaysia, Japan, Singapore, Finland, and Sweden) on the said topics, the author also aspires to promote the importance of including human perspectives in the transformation process of the automotive industry. With his knowledge of the back end of CV and AV systems, the author is also the secretary for the current SAE Cooperative Driving Automation Committee and regularly partakes in some of the SAE standardization committees. Furthermore, he also has distinctive experience in the product development area, in addition to R&D activities, with exposure to the four elements of ACES as the future mobility. This gives him a broad perspective on the ACES industry. The author understands that ACES is a very complex and broad subject that demands collaboration among different stakeholders; despite this—if executed properly—ACES will bring a lot of benefits to society (Chapter 7–9). Therefore, based on the author's experience above, he felt compelled to write a book that tries to explore this topic for the general audience to read without being too technical, while still adhering to and maintaining a few technological discussions. In other words, bringing the sky-high ACES discussions among the experts to the ground for the knowledge of the public audience. Thus, this book is born.

10.1. Recapitulating the Book

As this book is intended so that its chapters can be read as a stand-alone piece, this final chapter is then written to sum up the whole book. It is penned by providing comments for each section, i.e., Section 1 "Introduction," Section 2 "ACES as the Future Mobility: Background," Section 3 "Concise Overview of ACES," and Section 4 "Disruptions, Challenges, and Benefits of ACES."

In addition, the author also makes observations from several viewpoints on the required future efforts for ACES that need to be done. Consequently, this chapter consists of the book conclusion, the takeaway points, and potential "call for action" suggestions. Hopefully, the readers of Chapter 10 as a standalone chapter will also be interested to read the other chapters.

10.2. Summary for Section 1— Introduction

In Section 1, the author provides the background to the current industrial revolution. The discussions concisely touch on Industry 4.0, giving a glimpse of the future world where the arrival of a lot of new emerging technologies is inevitable. It is mentioned in Chapter 1 that the disruption will not only alter the technology but also change the future job landscape. It is apparent that the automotive industry is also impacted by these changes with the growth of ACES technology.

10.3. Summary for Section 2—ACES as the Future Mobility: Background

To assure that ACES as the future mobility is not just a mere daydream, the author did a comprehensive survey in Chapter 2 to understand the current trends that support the claim. The author then outlines a listing of 20 recent events and progress which propel the ACES growth. The trends are as follows:

1. Advanced Driver Assistance Systems and Vehicle Automation
2. Active Safety and the Safety Benefits
3. Software-Defined Vehicles
4. Fourth Industrial Revolution
5. United Nations Sustainable Development Goals
6. Society 5.0
7. Climate Change and the Intergovernmental Panel on Climate Change 2021 Report
8. Carbon-Neutral and Finite Petroleum Resources
9. Green, Circular, and Sustainable Economy
10. Sharing Economy
11. Regulations Facilitating ACES

12. Internet of Things and 5G
13. Advances in Computing Power and Platforms
14. Sensor Advancements (LiDAR, RADAR, Camera)
15. Cross-Functional Software Product Development and Change Management
16. Silicon Valley and Rise of Start-Ups
17. X-as-a-Service
18. Acqui-Hiring and M&A
19. Supply Chain Evolutions
20. Digital Natives

By reading this section, it is reassuring that the rise of ACES is supported by multidirectional trends and progress.

10.4. Summary for Section 3—Concise Overview of ACES

Section 3 consists of four chapters, i.e., Chapters 3, 4, 5, and 6, each providing a technological overview of autonomous vehicles (AVs), vehicle connectivity, vehicle electrification, and shared mobility.

In each of the chapters, in addition to the technical survey, the author also briefly mentions the current state of the industries. In this section, the readers, especially those with a non-technical background, will have a good understanding of the technical definition of the ACES elements.

10.5. Summary for Section 4— Disruptions, Challenges, and Benefits of ACES

One of the main distinguishing characteristics of this book is that, apart from providing an overview of ACES, it attempts to highlight how these innovations will touch our lives, us humans. Therefore, in Section 4, the disruptions, challenges, and benefits of ACES are reported.

Chapter 7 addresses the social, legal, technological, workforce, and ethical disruptions caused by ACES. The readers will be able to see that the productization of EVs not only brings a lower cost of ownership and economic benefits but also provides a cleaner environment for future generations.

Similarly, the discussions on the challenges and benefits of ACES are written in Chapters 8 and 9, respectively, from different viewpoints.

10.6. Required Future Efforts for ACES Mass Deployment

10.6.1. Technical

One of the most critical aspects for ACES to be a reliable and safe mobility solution is the requirements for it to be productized according to the regulation and required homologations. Rushing the implementation of ACES will not only bring bad product quality but might also cause fatalities. Therefore, awareness and propagation on the topics of ACES safety should be made continuously to various stakeholders.

Furthermore, for the technologies to be deployed worldwide including in developing countries and to assure a continuous good implementation of ACES technologies, the related technical curriculum should also be included in the education systems in countries worldwide. For example, to nurture programming talents, schools and universities around the world should provide the necessary basic programming skills to the younger generations. This will not only be helpful for ACES operations worldwide but also will solve the aforementioned "future of jobs availability" challenges.

10.6.2. Legal

In recent years, a lot of new regulations which are related to Industry 4.0, ACES, and future mobility are introduced. For example, the European Union Data Act [7], EU Battery Regulations

[8], and European Chips Act [9] are introduced or updated with a major direct impact on ACES mobility elements. However, more efforts need to be done on the perspectives of customer rights for ACES topics in different regions worldwide. As ACES will see the involvement of fully AD, ethical and safety questions such as "who will be blamed in the case of accidents of robotaxis" should also be discussed and regulated. Furthermore, topics such as insurances for ACES mobility in different use cases should also be expanded and introduced.

10.6.3. Social

It is not sufficient for AVs to be implemented and advanced only in developed countries. In fact the dire needs for ACES technologies are arguably in the less developed areas. It is the year 2022, but in some developing countries, some children are still using unsafe boats to cross the river to get to school because of a lack of infrastructure and public transportation [10, 11].

ACES has the potential to solve this issue. For example, if an AV is deployed in a rural area where it is challenging to station human drivers, it will help the said students get to school without having to commute in such a way (crossing the river in old boats). Furthermore, benefits can also be gained with AVs deployment in rural areas where buses are not available for elderly citizens to use. With safe ACES technology, transportation quality will be improved. However, this will require proper infrastructure construction. Therefore, collaborations among public institutions, governments, and private entities are needed to bring the best benefits of ACES to the public. Tax incentives can also be organized by governments to encourage the adoption of new technologies in their countries.

Efforts have been done by relevant bodies such as International Telecommunication Union (ITU) via the Focus Group on Artificial Intelligence for Autonomous and Assisted Driving (FG-AI4AD) to discuss AD developments and standardization. However, most of the discussions are still focusing on technical perspectives. Therefore, more efforts should be made to discuss the regulations of ACES implementation from the perspective of "societal impact." The author strongly encourages this kind of measure. As a note, if any parties are interested in initiating similar efforts, the author believes the automotive industry should be supportive of those movements [12].

10.6.4. Process and Procedures

As mentioned in several chapters of this book, for the acceptance of ACES technology to be widespread, the technology must be developed reliably. However, as ACES involve a lot of software development, a change in management philosophy and process might be needed within the automotive industry. These involve changes in the leadership style, service design with external and internal stakeholders, and adopting the "product" mindset for ACES technology development [13]. For this to happen, more transparent communication is needed within automotive organizations, where cross-functional collaborations should be encouraged [14].

10.7. It Is a Bumpy Road, But It Is Not Impossible

From the discussions in the book, similar to other disruptive technologies, it might seem like a bumpy road when looking at the future. However, an anecdote might help provide a different angle of perspective to this.

In 2017, the author of this book co-wrote a short paper called "Emerging Technologies with Disruptive Effects: A Review." It has since been read by more than 20,000 readers as of 2022 [15] (Figure 10.1). When the author co-wrote that paper five years ago, the

FIGURE 10.1 Snapshot of a paper that the author has co-written on the emerging technologies topic from the ResearchGate website [15].

widespread AV hype happening in the industry was still in its infancy, with no real clear direction. But when this part of the book was written and when the submissions of the book were partially done, a very interesting thing happened.

Ford announced the appointment of Joshua Sirefman, a former Alphabet leader, as the new Chief Executive Officer of Michigan Central, with the vision to be a center of "future mobility." This is a very interesting direction where the ACES developments are now involving not only automotive experts, but instead, the leadership of different fields is also imported to facilitate the full-picture implementation of ACES into society [16]. This shows that cross-organizational collaborations are starting to happen to enable the ACES big-picture to turn into reality. Several other similar ACES-motivated projects also emerged globally in various cities, including Dubai [17].

This shows that very interesting perspectives are growing in the industry. ACES as the future mobility is here to stay. The real implementation may be refined from time to time, but each of the elements in ACES, i.e., autonomous vehicles, vehicle connectivity, vehicle electrification, and shared mobility, is disrupting the auto-motive sector and will merge with other ecosystems in other indus-tries, which will improve our lives. We as a society have the collec-tive opportunity to define the ACES direction together for the future of our generation.

10.8. The Dream and Hope for This Book

The book has finally come to an end. The author hopes that this book will not just be another technical reference, but the author tried to write the book with some human touch and perspective. This is to remind the practitioners, engineers, industry, and business experts, among many others, that by the end of the day, all these ACES elements will be utilized by humans and our future generation.

Therefore, it is the author's hope for this book to be not just mere literature but yield a concrete action (help build a reliable

ACES implementation)—a book that can stimulate and build awareness. Accordingly, the readers are strongly encouraged to read the cited references throughout the chapters, and the author is confident that these will help them understand the topic better.

10.9. Summary and Conclusion

The writing of this book started in early 2021 and the journey has now come to an end. The author hopes that the readers of this chapter will read the whole book and eventually benefit from the discussions. Further collaborations with the author on ACES topics are welcome. If you find this book helpful, do feel free to spread the word about it and buy it for your library and office!

Toward ACES as the future mobility, we go!

References

1. IoT World Today, "Autonomous-Vehicle Software Startup Annotell Secures $24M Investment," accessed August 2022, https://www.iotworldtoday.com/2022/02/07/autonomous-vehicle-software-startup-annotell-secures-24m-investment

2. Hetzner, C., "Audi and Aptiv Invest $285 Million in 'Orchestral Director' for Self-Driving Car Data," *Fortune*, accessed August 2022, https://fortune.com/2022/02/03/audi-aptiv-invest-285-million-tttech-auto-self-driving-car-data

3. Evans, R., "Scania to Invest US$ 104M in New Test Track for Avs and EVs," Automotive Testing Technology International, accessed August 2022, https://www.automotivetestingtechnologyinternational.com/news/active-safety/scania-to-invest-us104m-in-new-test-track-for-avs-and-evs.html

4. ETAuto.com, "Hyundai Mobis to Invest up to USD 6.72 bn on Auto Chips, Robotics," *The Economic Times*, accessed August 2022, https://auto.economictimes.indiatimes.com/news/auto-components/hyundai-mobis-to-invest-up-to-usd-6-72-bn-on-auto-chips-robotics/89762023

5. TechCrunch, "Ibex Investors' Newest Fund Is Betting on a Mobility Revolution," accessed August 2022, https://techcrunch.com/2022/02/17/ibex-investors-newest-fund-is

6. Northvolt, "Volvo Cars and Northvolt to Open Gothenburg R&D Centre as Part of SEK 30bn Investment in Battery Development and Manufacturing," accessed February 2022, https://northvolt.com/articles/volvo-northvolt-dec2021/

7. European Commission, "Data Act: Commission Proposes Measures for a Fair and Innovative Data Economy," accessed August 2022, https://ec.europa.eu/commission/presscorner/detail/en/ip_22_1113

8. European Commission, "Batteries and Accumulators," accessed August 2022, https://ec.europa.eu/environment/topics/waste-and-recycling/batteries-and-accumulators_en

9. European Commission, "European Chips Act: Communication, Regulation, Joint Undertaking and Recommendation," accessed August 2022, https://digital-strategy.ec.europa.eu/en/library/european-chips-act-communication-regulation-joint-undertaking-and-recommendation

10. NBC News, "Risky River Crossing: Filipino Kids Tube to Get to School," accessed August 2022, https://www.nbcnews.com/news/photo/risky-river-crossing-filipino-kids-tube-get-school-flna1c6429149

11. BBC, "Getting Boats to Children Who Have to Swim to School," accessed August 2022, https://www.bbc.com/news/world-asia-47024828

12. ITU, "Focus Group on AI for Autonomous and Assisted Driving (FG-AI4AD)," accessed August 2022, https://www.itu.int/en/ITU-T/focusgroups/ai4ad/Pages/default.aspx

13. Wei, J., "Leading Change under Uncertainty: Lessons from a Multinational Automotive Company," 2020, 1-54, http://etds.lib.ntnu.edu.tw/cgi-bin/gs32/gsweb.cgi?o=dstdcdr&s=id=%22G060786030I%22.& http://rportal.lib.ntnu.edu.tw:80/handle/20.500.12235/109819

14. Hamid, U.Z.A., Irimescu, D.S., and Zaman, M.T., "Challenges of Complex Software Development for Emerging Technologies in Automotive Industry: Bridging the Gap of Knowledge Between the Industry Practitioners," SAE Technical Paper 2022-01-0109, 2022, https://doi.org/10.4271/2022-01-0109

15. Rahman, A., Airini, U.Z.A.H., and Chin, T.A., "Emerging Technologies with Disruptive Effects: A Review," *Perintis eJournal* 7, no. 2 (2017): 111-128.

16. Howard, P.W., "Ford Recruits CEO with Detroit Roots to Run Michigan Central Station, Corktown Site," accessed August 2022, https://eu.freep.com/story/money/cars/ford/2022/02/21/ford-recruits-michigan-central-station-ceo-nyc-detroit-roots/6877283001

17. Hafiz, D. and Zohdy, I., "The City Adaptation to the Autonomous Vehicles Implementation: Reimagining the Dubai City of Tomorrow," in Hamid, U.Z.A. and Al-Turjman, F. (eds.), *Towards Connected and Autonomous Vehicle Highways* (Cham: Springer, 2021), 27-41.

Index

A

Acqui-hiring, 30–31, 182
Active safety, 17, 74–75, 78, 138, 181
Adaptive Cruise Control, 17, 52
Advanced Driver Assistance Systems (ADAS), 16–17, 181
Airbnb Inc., 24, 108
Artificial Potential Field strategy, 51
Automotive SPICE, 160
Autonomous Emergency Braking (AEB), 17, 138, 162
Autonomous Mobile Robots (AMR), 42
Autonomous Vehicles (AVs), 3, 8, 24, 26, 182
  accidents, 150–151
  back-end algorithms, 43–44
  calibration, 56
  definitions, 41–42
  environmental awareness, 48–50
  human-operated vehicles operation, 44–46
  insurance industry, 130
  interfaces, 55–56
  mapping and localization, 47–48
  motion control, 54–55
  motion planning, 52–54
  production timeline, 19
  remote monitoring, 56
  risk assessment, 50–52
  robotaxi, 172
  software architecture, 46
  vision, 56–58
Autonomous Vehicles Readiness Index, 56–57

B

Battery Electric Vehicles (BEVs), 94–98, 100
Battery management and recycling, 148–149
Business-facing organizations, 152–153

C

Calibration, 56
California Consumer Privacy Act, 147
Carbon emissions, 20, 22, 91
Carbon-neutral, 22, 181
Carpooling, 112–113, 173
Carsharing, 112
CarSim, 54
Cellular Vehicle-to-Everything (C-V2X), 70–71
Change management, 28, 154, 182
Circular economy, 23, 139, 181
Climate Action, 91
Climate change, 20–22, 90–91, 139, 181
Cloud technology, 69

Computing power and platforms, 25–27
Connected and Automated Mobility (CAM), 24
Connected Vehicles (CVs), 8, 182
  active safety, 74–75
  automated delivery, 77
  benefits, 66–67
  capabilities, 67
  cloud technology, 69
  C-V2X, 70–71
  development, 19
  DSRC, 69–70
  elements/methodology, 67
  FIR, 19
  5G, 71–72, 128
  GPS, 76–77
  hit-and-run crimes, 167–169
  improved infotainment, 76
  Internet growth, 66
  LPWAN, 71
  roles and contributions, 78–79
  safety, 17, 162
  shared mobility, 77
  6G, 72–73
  Smart City, 66
  smartphones, 65–66
  traffic jam reduction, 76–77
  VANET, 68–69
  vehicle platooning, 75–76
  Vehicle-to-Everything, 73–74
  vision, 79–80

Coronavirus disease of 2019 (COVID-19), 32, 148–149
Cross-functional software product development, 28, 182
Customer-and-user-facing technology development, 116–117
Cybersecurity, 147, 163–164

D

Data privacy, 69, 134
Dedicated Short-Range Communications (DSRC), 69–70, 73–74
Digital Natives, 31–32, 182
Disruptive innovation, 135
  business models, 136
  climate change issues, 139
  democratizing mobility, 138
  income for countries, 138
  industrial stakeholders, 138
  infrastructural disruptions, 137
  legal, economical, and workforce perspectives entertainment industry, 132–133
  insurance industry, 130
  job ecosystems opportunities, 132
  leadership requirements, 135
  ownership of vehicles, 132
  passenger behavior, journey, 133
  private-public-people partnerships, 134–135
  regulations, 133–134
  taxations, 130–131
  workforce skills, 131

process and regulations changes, 136–137
Skunkworks project-organizations, 138–139
social disruption (*see* Social disruption)
software-defined vehicles, 136

E

Economic benefits business models, 173
  cross-collaborations, 174
  investments potential, 172–173
  opportunities for developing countries, 173
"Economies of scale," 151–152
Edge case scenarios, 150
Education, 154
Electric motors (EMs), 93–95
Electric Vehicles (EVs), 8, 182
  battery recycling and waste management, 148–149
  BEVs, 96–98
  carmakers, 90
  charging facilities, 148
  companies, 31
  electrification-themed development, 89
  environmental issues, 90–92
  FCEVs, 98–100
  Generation Z, 88–89
  HEVs, 92–95
  ICE, 87–88
  infrastructural improvements, 148
  Millennials, 88–89
  PHEVs, 95–96
  range anxiety, 147–148
  roles, 101
  Solar EVs, 100
  vision, 101–102

Emergency Lane Keeping, 17
"Emerging Technologies with Disruptive Effects: A Review," 185–186
Environmental awareness system, 48–50
EU Battery Regulations, 183
European Chips Act, 184
European Union Agency for Cybersecurity (ENISA), 24
European Union Data Act, 134, 147, 183

F

Fifth-generation (5G) technology, 4–5, 67, 182
  C-V2X, 70
  growth, 24–25
  for vehicle connectivity, 71–72
Focus Group on Artificial Intelligence for Autonomous and Assisted Driving (FG-AI4AD), 184
Fourth Industrial Revolution (FIR), 3–4, 7, 19–20, 23, 29, 66, 109, 181
Fuel Cell Electric Vehicles (FCEVs), 98–100

G

Generation Z, 88–89
GPS, 76–77
Green economy, 23, 181
Greenwashing, 155

H

Hybrid Electric Vehicles (HEVs), 92–95

I

Industry 4.0. *See* Fourth Industrial Revolution (FIR)
Insurance industry, 130
Interfaces, 55–56

Intergovernmental Panel
on Climate Change
(IPCC) 2021,
20–22, 181
Internal Combustion
Engine (ICE), 22,
87–88, 93–94, 98
International
Telecommunication
Union (ITU), 184
Internet, 75, 108, 110–111
Internet of Things (IoT), 19,
24–25, 66, 182
Internet of Vehicle (IoV), 25
In-vehicle monitoring, 154
IPG CarMaker, 54

J
Job landscapes, 153, 181

L
Light Detection and
Ranging (LiDAR),
26–27, 46, 48
Lightyear One, 100
Low-Power Wide-Area
Network
(LPWAN), 71
LuxTurrim5G, 171–172

M
Maritime autonomous
surface vessels, 42
Mass deployment
legal, 183–184
process and procedures,
185
social, 184–185
technical, 183
Mergers and acquisitions
(M&A), 6, 31, 182
Micromobility, 114–115
Microtransit, 115
Millennials, 88–89
Misleading marketing,
154–155
Mobility-as-a-Service
(MaaS), 110
Mobility user benefits,
171–172
Motion control, 54–55
Motion planning, 52–54

N
Network latency, 147

O
Offshore wind farms, 90–91

P
Paratransit, 115
Petroleum resources,
22, 181
Petty crimes, 150, 154–155
Plug-In Hybrid Electric
Vehicles (PHEVs),
95–96
Pricing, 151–152
Privacy, 70, 79, 80, 163–164

R
Radio Detection and
Ranging (RADAR),
26–27, 46, 48
Radio Frequency
Identification
(RFID), 66
Range anxiety, 90, 94, 97,
147–148
Regenerative braking,
93, 100
Regulations, 24, 181,
151, 153
Remote monitoring, 56
Research and development
(R&D), 25–26,
179–180
Ridehailing, 114, 163
Ridesharing, 113, 149–150
Ridesourcing, 114
Risk assessment, 50–53
Robotaxi, 112, 126,
163–164, 172

S
Safety benefits, 181
transportation domains,
160–162
of vehicular automation,
160–162
vehicle connectivity,
162–163
Sanitation management,
148–149
Security benefits, 163–164

Sensors, 47, 56, 71,
78–79, 182
cameras, 26–27, 46, 48
LiDAR, 26–27, 46, 48
RADAR, 26–27, 46, 48
Shared mobility, 4–6, 8,
19, 182
business models
development,
116–117
carpooling, 112–113
carsharing, 112
definition, 111–112
micromobility, 114–115
microtransit, 115
paratransit, 115
pricing, 151–152
private vehicle
ownerships,
164–165
ridehailing, 114
ridesharing, 113
ridesourcing, 114
roles, 117–119
safety, 78
sanitation management,
148–149
sharing economy,
108–109
stakeholders, 119
vs. traditional taxi, 116
traffic jams in cities,
167–168
uberization, 110
vehicle connectivity, 77
XaaS, 110–111
Sharing economy, 19,
23–24, 181
Silicon Valley, 28–29, 182
Sixth-generation (6G)
technology, 72–73
Social disruption
autonomous VTOL
technology, 128
blockchain
technology, 128
media and infotainment,
126–127
mobility, 128
overview, 125–126

urban planning, 128–129
work-life balance improvements, 126
Society 5.0, 7, 20, 181
Software-defined vehicles, 18–19, 69, 181
Software development, 28, 151
Solar-based EVs (SEVs), 100
Stakeholders, 119, 130, 154–155
Standardizations, 24, 153, 160, 184
Start-ups, 23, 28–29, 31, 88, 98, 182
Supply chain evolutions, 31, 182
Sustainable economy, 23, 181
Sustainable mobility ecosystem
   air quality improvement, 170–171
   crimes reduction, 167–169
   health quality improvement, 170–171
   improved traffic in cities, 165–166
   low vehicle emission, 165–166
   shared mobility, 167–168
   social security benefits, 168–169
   UN SDGs, 169–170
   urban recreational zones and activities, 166–167

T

Taxations, 130–131
Technological benefits, 163–164
Traffic jam, 149–150
Transportation domains, 160–162
Transportation Network Company (TNC), 114
Traveling salesman problem, 150

U

Uberization, 29, 110
Uncertainties, 146–147, 150
2021 UN Climate Change Conference (COP26), 92
United Nations Sustainable Development Goals (UN SDGs), 19–20, 78, 169–170, 181

V

Vandalism, 150, 154–155
Vehicle automation, 4–6, 16–17. *See also* Autonomous Vehicles (AVs)
Vehicle connectivity, 4–6. *See also* Connected Vehicles (CVs)
Vehicle electrification, 4–6, 89. *See also* Electric Vehicles (EVs)
Vehicle ownership, reduction in, 164–165

Vehicle safety, 16, 181
   technologies, 17
   transportation domains, 160–162
   vehicle connectivity, 162–163
   of vehicular automation, 160–162
Vehicle-to-Everything (V2X), 25, 73–74, 137
Vehicle-to-roadside communication, 68, 69
Vehicle-to-Vehicle (V2V) communication, 68, 69
Vehicular ad hoc network (VANET), 68–69
Vertical Take-Off and Landing (VTOL) aircraft, 42
Visionary leadership, 154
Vulnerable road users (VRU), 50

W

Wide-Area Network (WAN), 71
Workforce skills, 131

X

X-as-a-Service (XaaS), 29–30, 110–111, 182

Z

Zero-emission vehicles, 92

About the Author

Umar Zakir Abdul Hamid, PhD, has been working in the future mobility (connected and autonomous vehicle) field with various teams in different countries and continents since 2014. Previously, he led a team of 12 engineers (of 10 different nationalities) working on the Autonomous Vehicle Software Product Development at Sensible  4, Finland. Umar is one of the recipients of the Finnish Engineering Award 2020 for his contributions to the development of all-weather autonomous driving solutions with the said company. With more than 30 scientific publications as an author and editor under his belt, Umar actively participates in global automotive standardization efforts where he is a Secretary of the SAE Cooperative Driving Automation Committee. Since the end of the summer of 2021, Umar has been working as the Lead of Strategic Planning for CEVT AB in Sweden. He is also currently holding the position of Secretary for IEEE Sweden. Umar believes that the automotive and mobility industry is facing massive disruption; therefore, the transformation needs to be understood by all of the concerned stakeholders in the ecosystem. This book is written to share a general knowledge on Autonomous, Connected, Electric, and Shared (ACES) Mobility to the public audience as well as aspiring researchers and experts. The author believes that his expertise, knowledge, and experience during the years in the said field will be useful to provide practical insights to the potential readers on the expressed themes.

Autonomous, Connected, Electric and Shared Vehicles

Disrupting the Automotive and Mobility Sectors

Umar Zakir Abdul Hamid

We are at the beginning of the next major disruptive cycle caused by computing. In transportation, the term **Autonomous, Connected, Electric, and Shared (ACES)** has been coined to represent the enormous innovations enabled by underlying electronics technology. The benefits of ACES vehicles range from improved safety, reduced congestion, and lower stress for car occupants to social inclusion, lower emissions, and better road utilization due to optimal integration of private and public transport.

ACES is creating a new automotive and industrial ecosystem that will disrupt not only the technical development of transportation but also the management and supply chain of the industry. Disruptions caused by ACES are prompted by not only technology but also by a shift from a traditional to a software-based mindset, embodied by the arrival of a new generation of automotive industry workforce.

In **Autonomous, Connected, Electric and Shared Vehicles: Disrupting the Automotive and Mobility Sectors**, Umar Zakir Abdul Hamid provides an overview of ACES technology for cross-disciplinary audiences, including researchers, academics, and automotive professionals. Hamid bridges the gap among the book's varied audiences, exploring the development and deployment of ACES vehicles and the disruptions, challenges, and potential benefits of this new technology.

Topics covered include:

- Recent trends and progress stimulating ACES growth and development
- ACES vehicle overview
- Automotive and mobility industry disruptions caused by ACES
- Challenges of ACES implementation
- Potential benefits of the ACES ecosystem

While market introduction of ACES vehicles that are fully automated and capable of unsupervised driving in an unstructured environment is still a long-term goal, the future of mobility will be ACES, and the transportation industry must prepare for this transition. Autonomous, Connected, Electric and Shared Vehicles is a necessary resource for anyone interested in the successful and reliable implementation of ACES.

Cover image used under license from Shutterstock.com

RELATED RESOURCES:

ADAS and Automated Driving: A Practical Approach to Verification and Validation
Plato Pathrose
978-1-4686-0412-2

Fundamentals of Connected and Automated Vehicles
Jeffery Wishart
978-0-7680-9980-5

ISBN: 978-1-4686-0347-7

INTERNATIONAL®

9 781468 603477

Printed in the USA
CPSIA information can be obtained
at www.ICGtesting.com
LVHW071934121124
796372LV00008B/33